The Mata Book

A Book for Serious Programmers and Those Who Want to Be

The Mata Book

A Book for Serious Programmers and Those Who Want to Be

William W. Gould
STATACORP LLC

A Stata Press Publication
StataCorp LLC
College Station, Texas

Published by Stata Press, 4905 Lakeway Drive, College Station, Texas 77845
Typeset in LATEX 2_ε
Printed in the United States of America

10 9 8 7 6 5 4 3 2 1

Print ISBN-10: 1-59718-263-X
Print ISBN-13: 978-1-59718-263-8
ePub ISBN-10: 1-59718-264-8
ePub ISBN-13: 978-1-59718-264-5
Mobi ISBN-10: 1-59718-265-6
Mobi ISBN-13: 978-1-59718-265-2

Library of Congress Control Number: 2018933411

Contents

Acknowledgment

I thank Nicholas J. Cox for suffering through an earlier draft of this book and identifying the places where I was pedantic, abstruse, or off on a tangent. I thought I was done but Nick convinced me otherwise. This book is better because of him. Remaining inelegancies and errors are mine.

1 Introduction

1.1 Is this book for me?

This book is for you if you have tried to learn Mata by reading the *Mata Reference Manual* and failed. You are not alone. Though the manual describes the parts of Mata, it never gets around to telling you what Mata is, what is special about Mata, what you might do with Mata, or even how Mata's parts fit together. This book does that.

This is an applied book. It will teach you the modern way to write programs, which is to say, it will teach you about structures, classes, and pointers. And the book will show you some programming techniques that may be new to you. In short, in this book, we are going to use Mata to write programs that are good enough that StataCorp could distribute them.

This book is for "serious programmers and those who want to be". Fifteen years ago, the subtitle would have referenced professional rather than serious programmers, and yet I would have written the same book. These days, the distinction is evaporating. I meet researchers who do not program for a living but are most certainly serious. And I meet the other kind, too.

A serious programmer is someone who has a serious interest in sharpening their programming skills and broadening their knowledge of programming tools. There is an easy test to determine whether you are serious. If I tell you that I know of a new technique for programming interrelated equations and your response is "Tell me about it," then you are serious.

Being serious is a matter of attitude, not current skill level or knowledge.

Still, I made assumptions in writing this book. I assumed that you have some experience with at least one programming language, be it Stata's ado, Python, Java, C++, Fortran, or any other language you care to mention. I also assumed that you already know that programs contain conditional statements and loops. If you need a first introduction to

1

programming, you could look at the introductory section of the Mata manual or at the Mata chapters in Baum's friendly text *An Introduction to Stata Programming* (2016).

The examples in this book are statistical and mathematical. Formulas are provided, but the formulas are of secondary importance. They just provide the examples of something for us to program.

In this book, I will show you a language aimed at programming statistical and data management applications that has all the usual features and some unique ones, too. And I will show you programming techniques that might be new to you.

As I said, being serious is a matter of attitude. New techniques and languages are continually being developed, and you need to learn them, just as I still learn them. I have been programming for 45 years as a professional. I have a lot of experience and knowledge, but I have not stopped learning new techniques. I may be a professional programmer, but more importantly, I am a serious one.

1.2   What is Mata?

Many Stata users would describe Mata as a matrix language. StataCorp itself markets Mata that way. Mata would be more accurately described, however, as an across-platform portable-code compiled programming language that happens to have matrix capabilities. Just as important as its matrix capabilities are Mata's structures, classes, and pointers.

We at StataCorp designed and wrote Mata to be the development language that we would use. Nowadays, we write most new features of Stata in Mata. Before Mata existed, we used C. Compared with C, Mata code is easier to write, less error prone, easier to debug, and easier to maintain.

It is important that Mata is compiled. Being compiled means that programs run fast. Stata's other programming language, ado, is interpreted. Interpreted languages are slow in comparison with compiled languages. Mata code runs 10–40 times faster than ado.

Mata looks a lot like C and C++. In *The C Programming Language*, Kernighan and Ritchie (1978) introduced what has become perhaps the most famous first program:

```
main()
{
        printf("hello, world\n") ;
}
```

To convert the program to Mata, we need to add `void` in front of `main()`:

```
: void main()
> {
>         printf("hello, world\n") ;
> }
: main()
hello, world
```

Most Mata users would not bother typing the semicolon at the end of `printf("hello, world\n")`. Semicolons are optional in Mata. There are other differences between the languages, too. Those differences are covered in appendix C.

1.3   What is covered in this book

The programs we will write in this book are

Filename	Contents
hello.mata	First program, function `hello()`
n_choose_k.mata	Serious but short function, packaged as library function
lr1.mata	Linear regression, ver. 1 (structures)
lr2.mata	Linear regression, ver. 2 (structures)
earthdistance.mata	An aside concerning classes
linreg1.mata	Linear regression take 2, ver. 1 (classes)
linreg2.mata	Linear regression take 2, ver. 2 (classes)
spmat1.mata	Sparse matrices, ver. 1
spmat2.mata	Sparse matrices, ver. 2
spmat3.mata	Sparse matrices, ver. 3

The first serious program we will write is `n_choose_k()`. It will have just 47 lines including comments and white space.

We will then work our way to a nearly complete implementation of linear regression, starting with `lr1.mata` and ending with `linreg2.mata`. There will be only 388 lines in the final code in `linreg2.mata`! We will use structures for the first two implementations and use classes after that.

The `earthdistance.mata` program merely illustrates a point about class programming.

Finally, we will undertake a large project, namely, the implementation of sparse matrices. Sparse matrices are matrices in which most elements are 0. The project will concern storing the matrices efficiently—there is no reason to store all those 0s—and writing code to add and multiply them just as if they were regular matrices. File `spmat3.mata` will contain 937 lines.

We will do all that, but we will not start until chapter 9. There is a lot to tell you first.

Chapter 2 covers the mechanics of using Mata. You may know that Mata can be used interactively, but that is not how we will be using it except when we want to experiment before committing an idea to code.

Chapter 3 takes you on a tour of Mata. It will show you ordinary features, such as assignment; surprising features, such as 0×0 matrices and 0×1 and 1×0 vectors; and advanced features, such as structures, classes, and pointers. Pointers, by the way, are not nearly as difficult to understand as you might fear. Later, we will use pointers when we write `lr1.mata`, our first implementation of linear regression, and we will use them in an advanced way when we write `spmat3.mata` to implement sparse matrices.

Chapter 4 explains Mata's programming statements, all nine of them. There may be only nine, but they fit together in remarkable ways.

Chapter 5 provides details about Mata's expressions, such as `y = sqrt(2)`. Expressions are one of the nine programming statements, but that understates their importance because they comprise the bulk of programs. Just calling a subroutine is an expression. Chapter 5 also discusses programming for numerical accuracy. Do not skip section 5.2.1.2 even though its title is *Base-2 notation*.

Chapter 6 describes Mata's 40 variable types. One of them is `transmorphic`, and the chapter enumerates its proper and improper uses.

Chapter 7 is about Mata's `strict` option. `strict` tells Mata to flag questionable constructs in programs. Bugs hide inside questionable constructs.

Chapter 8 is about function arguments. Mata passes arguments by reference, but you may not yet know what that means. The chapter also shows how to write functions that allow a varying number of arguments.

In chapter 9, we finally turn to programming. The chapter is entitled *n_choose_k() three ways*. We will write the new function `n_choose_k()` and use it in three ways. We will use the function in an analysis do-file, as the computational engine inside an ado-file, and as a function to be added to a Mata library so that it can be used anywhere and anyplace.

We will start programming in chapter 9, and we will not stop. A few chapters after 9 will explain Mata features that we will need for the programs we will write. Chapters 10 and 11 deeply explain structures. Chapters 12 and 13 do the same for classes. Chapter 14 shows how to create new variable types so you can declare a variable to be `boolean` instead of `real` or an `SpMat` instead of a `class SpMat scalar`. Chapter 15 shows a better way to deal with constants that appear in code. Chapter 16 explains Mata's associative arrays.

The chapters of this book are about Mata, not Stata. All but one example is about writing Mata programs to be called from other Mata programs. And yet, the purpose of Mata is to add new features to Stata. In appendix A, we will finally discuss programming for Stata. Because you will have read the chapters, we will be able to discuss the subject as one serious and knowledgeable programmer with another. There will be three issues for us to discuss.

The first issue is how code should be structured. Stata's ado language is how new commands are added to Stata, and Mata does not change that. The question is whether you should write one line of ado-code calling Mata so that the entire program is written in Mata, or you should parse in Stata and then call Mata, or you should leave the ado-code in charge and use Mata to provide the occasional subroutine for the ado-code to call.

The second issue is how to access Stata objects such as variables, observations, macros, and the like. Mata provides functions to do this.

The third issue is how to handle errors caused by mistakes by the users of our code. By default, Mata aborts with error and issues a traceback log. That is acceptable behavior when we write subroutines for use by other serious programmers, but it is not acceptable when writing code for direct use by Stata users. Mata has functions that will issue informative error messages and stop execution with a nonzero return code so that we can write code that handles errors as gracefully as Stata users expect.

The book covers more, too. A thorough treatment of programming requires discussion about workflow. Workflow is jargon for how to organize your work from the time you write the first line of code to the time the program is ready to ship or be put in use. Workflow is also about how you will later fix the program's first reported bug, and its second, and the substantive expansion of capabilities that you will make two years from now.

The workflow discussion begins in chapter 2, becomes more detailed in chapter 9, and continues in every programming example thereafter. Earlier, I mentioned the programs we will be writing: `hello.mata`, `n_choose_k.mata`, `lr1.mata`, and so on. When we write `lr1.mata`, we will also write file `test_lr1.do`, a Stata do-file to certify that the code in `lr1.mata` produces correct results. We will store the certified code and its test file in our Approved Source Directory. We will develop an automated procedure for creating and updating Mata libraries that recompiles all the code in all the `*.mata` files, runs all the `test_*.do` files, and rebuilds libraries from scratch.

In this book, we will produce not only programs, such as `hello.mata`, `n_choose_k.mata`, `lr1.mata`, and others, but also their workflow files. Here is the full set of files we will produce:

Filename	Contents
`hello.mata`	First program, function `hello()`
`n_choose_k.do`	Serious but short program, packaged as do-file
`nchooseki.ado`	Same function, packaged as ado-file
`test_nchooseki.do`	Certify that `nchooseki.ado` works
`n_choose_k.mata`	Same function, packaged as library function
`test_n_choose_k.do`	Certify that `n_choose_k.mata` works
`lr1.mata`	Linear regression, ver. 1 (structures)
`test_lr1.do`	Certify that `lr1.mata` works
`lr2.mata`	Linear regression, ver. 2 (structures)
`test_lr2.do`	Certify that `lr2.mata` works
`earthdistance.mata`	An aside concerning classes
`test_earthdistance.do`	Certify that `earthdistance.mata` works
`linreg1.mata`	Linear regression take 2, ver. 1 (classes)
`test_linreg1.do`	Certify that `linreg1.mata` works
`linreg2.mata`	Linear regression take 2, ver. 2 (classes)
`test_linreg2.do`	Certify that `linreg2.mata` works
`spmat1.mata`	Sparse matrices, ver. 1
`test_spmat1.do`	Certify that `spmat1.mata` works
`spmat2.mata`	Sparse matrices, ver. 2
`test_spmat2.do`	Certify that `spmat2.mata` works
`spmat3.mata`	Sparse matrices, ver. 3
`test_spmat3.do`	Certify that `spmat3.mata` works
`test.do`	Run all `test_*.do` certification files
`make_lmatabook.do`	Place all functions in `*.mata` files in Mata library

1.4   How to download the files for this book

Point your browser to

```
http://www.stata-press.com/data/tmb.html
```

I recommend that you download the files to a new, empty folder named `~/matabook/`. Then, if you want to look at `hello.mata`, you can type

```
. view ~/matabook/hello.mata
```

Note for Windows users: Type the above just as shown. Stata for Windows understands that / means \ and that ~ means your home directory.

You can download the files using Stata by typing

```
. mkdir ~/matabook/
. cd ~/matabook/
. copy http://www.stata-press.com/data/tmb/tmb.zip .
. unzipfile tmb.zip
. erase tmb.zip
. view README.txt
```

2 The mechanics of using Mata

2.1 Introduction

I showed the Mata function for `hello()` in the last chapter. Here it is again, although this time I have changed the function's name from `main()` to `hello()` and I execute it:

```
. mata:
                                        mata (type end to exit)
: void hello()
> {
>          printf("hello, world\n") ;
> }
: hello()
hello, world
: end
```

```
. _
```

Just the act of entering the program caused Mata to compile it. Mata compiled `hello()`, discarded the original source code, and left the compiled code in memory. That is why I can execute the function by typing `hello()`.

This interactive approach can be useful in teaching, but it is useless for serious applications. There are three ways Mata code is used more seriously.

Mata code can be placed in do-files. The functions you define there can be used interactively and by other do-files.

Mata code can be placed in ado-files. The functions you define there can be used inside the ado-file.

Mata code can be compiled and placed in libraries. The functions you place in them may be used anytime, anywhere. They can be used interactively, in do-files, in ado-files, and in other functions that appear in the same or different libraries.

2.2   Mata code appearing in do-files

I do not recommend putting Mata code straight into analysis do-files, although I have done that when the code was simple enough. Complicated code will need debugging, and debugging is easier when the code can be worked on in isolation. That argues for putting the code in its own do-file. Doing that also makes it easier to use the Mata code in other analyses.

I recommend that you place the code in its own do-file with the file extension `.mata`, such as

```
                                              ─────── hello.mata ──────
version 15

mata:
void hello()
{
        printf("hello, world\n")
}
end
                                              ─────── hello.mata ──────
```

Additional functions can appear in the same file:

```
                                              ─────── hello.mata ──────
version 15

mata:
void hello()
{
        printf("hello, world\n")
}
void goodbye()
{
        printf("good-bye, world\n")
}
end
                                              ─────── hello.mata ──────
```

Functions in the same file should be related. `hello()` and `goodbye()` are related; in the unlikely event you want to use one of them, you will probably need the other. Related use is a fine reason for functions to appear in the same file. Usually, however, the functions are even more related in that they call one another.

To use the functions in your analysis do-file, code do *filename*.mata in the do-file before using them:

```
─────────────────────────────────────────────── analysis.do ──────────
version 15
clear all
.
.
.
do hello.mata
mata: hello()
mata: goodbye()
.
.
.
─────────────────────────────────────────────── analysis.do ──────────
```

The line `do hello.mata` appears in boldface only for emphasis. When the `analysis.do` do-file executes the line, the `hello.mata` do-file will be executed, which will define the functions `hello()` and `goodbye()`. You could execute `analysis.do` by typing

```
. do analysis
  (output omitted )
```

File `analysis.do` begins with the line **version 15**. Version control is a hallmark of Stata. Every Stata do-file and ado-file since Stata 1 (in 1985) still works even though Stata's programming language looks nothing like it did originally. I included **version 15** in this file so that it will continue to work in the future.

In this book, we will use .mata for files containing Mata code. Those files should start with a version number, too. Look back and you will see that **version 15** appears at the top of file `hello.mata`. Version control serves the same purpose in .mata files that it does in do-files and ado-files. If some Mata language feature should change in the future, that feature will be backdated to have its old meaning. The version number does not preclude the use of features added later; it merely handles backdating for changes in syntax.

If there seems to be a profusion of **version 15** statements in these two files, imagine that it is now two-and-a-half years later and you are using Stata 16. I also need you to imagine that `analysis.do` is a real analysis do-file and that `hello()` and `goodbye()` do something useful. Typing `do analysis` will obviously reproduce the original results, but that is not what you want to do. You want to add a second analysis using a new Stata 16 feature. You create file `analysis2.do` and it starts, naturally enough, with **version 16**. You also want to use `hello()` and `goodbye()` in the new file, so you include `do hello.mata` in new file `analysis2.do`. The **version 15** in `mata.do` will assure that the code in `hello()` and `goodbye()` is given the Stata 15 interpretation when the functions are compiled. Thus, even in this new Stata 16 do-file, old functions `hello()` and `goodbye()` will work as originally intended.

2.3   Mata code appearing in ado-files

Mata code can appear in ado-files along with the usual ado-language code. The Mata code appears at the bottom of the file:

```
                                                          hello.ado
*! version 1.0.0  24 January 2018
program hello
        version 15
            .
            .
        mata: hello()
            .
            .
        mata: goodbye()
            .
            .
end
version 15
mata:
void hello()
{
        printf("hello, world\n")
}
void goodbye()
{
        printf("good-bye, world\n")
}
end
                                                          hello.ado
```

Mata code that appears in ado-files is treated specially. Rather than creating functions that anyone can use, the functions are made private. The ado-file can use them, but outside of the file, the functions do not even exist. You can name the functions as you please, even if those names are being used elsewhere.

There are three version numbers in the ado-file:

```
*! version 1.0.0  24 January 2018
program hello
    version 15
        .
        .
end
version 15
mata:
    .
    .
end
```

The first is merely a comment. It is how we at StataCorp track ado-file revisions. You should do something similar.

The second is the version to be used to interpret the ado-language code. It appears inside `program hello`. The version will take effect when the lines are executed. Ado-language lines are interpreted as they are executed.

The third version number is the version to be used to compile the Mata code. The version number could be different from the version number inside `program hello` (although it will usually be the same because the ado-code and its Mata subroutines will have been written at same time).

So much for version numbers. Ado-files can be difficult to debug because they have their subroutines. That means that you cannot interactively try functions `hello()` and `goodbye()` to test whether they are working as they should. If you need to debug `hello.ado`, the solution is to treat the ado-file as if it were a do-file by typing `do` *filename*. You type

```
. clear all
. do hello.ado
```

Typing `do hello.ado` causes `hello.ado` to be run just as any do-file would be run, and that means all the programs and functions defined in the file will be public. That also means you can interactively use `hello()` and `goodbye()` to verify that they are working properly:

```
. clear all
. do hello.ado
  (output omitted)
end of do-file
. mata:
———————————————————————————————— mata (type end to exit) ————————
: hello()
hello, world

: goodbye()
goodbye, world

: end
————————————————————————————————————————————————————————————————

. _
```

I usually create a do-file to accompany my ado-files while I am developing them.

```
──────────────────────────────────────────── hello.do ───────────
clear all

do hello.ado

mata:
hello()
goodbye()
end

  .

  .

  .
──────────────────────────────────────────── hello.do ───────────
```

File `hello.do` makes testing `hello.ado` easier. I can even place lines to test the code
in the do-file. Notice that in the above file, I directly tested Mata functions `hello()`
and `goodbye()`. Those functions would have been unreachable had I let the ado-file
load itself.

2.4   Mata code to be exposed publicly

In the two sections above, we created functions `hello()` and `goodbye()` for our private
use. In the first case, we created the functions for use in do-files in which we would
include the line `do hello.mata` before we used them. In the second case, we included
the functions inside an ado-file for the ado-file's exclusive use.

Mata also has public functions. Mata's built-in `sqrt()` function for calculating square
roots is public. It can be used in any program that we write. We do not have to load
it first, and the function is not just for the use of this program or that. You can create
public functions. Functions `hello()` and `goodbye()` could be public. Functions are
public in Mata when their compiled code is stored in a Mata library.

You develop the code for public functions just as you would for functions to be used
in a do-file, which is by storing their code in a `.mata` file. We previously created file
`hello.mata` containing the code for `hello()` and `goodbye()`:

```
──────────────────────────────────────────── hello.mata ───────────
version 15

mata:
void hello()
{
        printf("hello, world\n")
}
void goodbye()
{
        printf("good-bye, world\n")
}
end
──────────────────────────────────────────── hello.mata ───────────
```

All we need to do to make functions `hello()` and `goodbye()` public is compile the code and save it in a library. To compile the code, we just need to type `do hello` because, when Mata reads code, it compiles it and stores the result in memory.

To save the compiled code in a library, we use the `lmbuild` *libraryname* command. Library names must begin with the letter `l` (that is, a lowercase L) and end in `.mlib`. Here are a few valid library names:

```
lpolynomial.mlib
lstat.mlib
lxyz.mlib
```

Names that do not start with the letter `l` are invalid library names. We could not name a library `matabook.mlib`, but we could name it `lmatabook.mlib`. That is exactly what we are going to name our library. We are going to store all the functions we develop in this book in library `lmatabook.mlib`, starting with the functions `hello()` and `goodbye()`. The functions are stored in file `hello.mata`. To compile the functions and build library `lmatabook.mlib` containing them, we type

```
. clear all

. do hello.mata

. lmbuild lmatabook.mlib
  (output omitted)
```

The `clear all` command cleared Mata. The `do hello.mata` command compiled the functions. The `lmbuild lmatabook.mlib` command stored the compiled code (or all functions stored in memory) in `lmatabook.mlib`. The result is that `hello()` and `goodbye()` are now part of Mata. If you typed `clear all` or if you exited Stata and restarted it, you would discover that the functions still work:

```
. mata:
─────────────────────────── mata (type end to exit) ───────────
: hello()
hello, world

: goodbye()
good-bye, world

: end
─────────────────────────────────────────────────────────────

. _
```

If you modify the source code in `hello.mata`, you must rebuild the library:

```
. clear all

. do hello.mata

. lmbuild lmatabook.mlib, replace
  (output omitted)
```

You will reduce mistakes if you create a do-file to create the library:

```
————————————————————————— make_lmatabook.do ——————
// version number intentionally omitted

clear all
do hello.mata

lmbuild lmatabook.mlib, replace
————————————————————————— make_lmatabook.do ——————
```

Now to create or re-create the library, you can type

```
. do make_lmatabook
  (output omitted )
```

You can add additional do *name*.mata lines to make_lmatabook.do as you write them. Libraries can hold 1,024 functions, although you can increase that to 2,048 by specifying option size(2048) at the end of lmbuild.

3   A programmer's tour of Mata

3.1   Preliminaries

Before we start in earnest, let me show you around. As with all tours, I am going to show you some things you will not understand, or perhaps you will not understand why they are important. Just walk on by. We will return to the topics in later chapters.

Mata has the usual features you would expect, and it has some unexpected ones. It is hardly surprising that Mata can assign values to variables, such as

```
a = 0
```

17

Perhaps surprisingly, Mata also allows you to assign values to multiple variables at the same time:

```
a = b = c = 0
```

Mata even allows

```
a = ((b=c)+1)
```

The above line sets b equal to c and sets a equal to b+1.

It is also not surprising that Mata has vectors and matrices, but it may be surprising that they can have zero rows, zero columns, or both.

We will use Mata's interactive mode during the tour. You have already seen it in use:

```
.                                         // we are in Stata
. mata                                    // we enter Mata
                                       ─── mata (type end to exit) ───
: void hello()
> {
>          printf("hello, world\n")
> }
: hello()
hello, world
: end                          // we exit Mata

.  _                           // we are back in Stata
```

That Mata even has an interactive mode is likely to be a surprise because most compilers do not. Entering a program interactively, even one as simple as hello(), is not something we will be doing in later chapters. Interactive mode is nonetheless useful for experimenting with functions and features before using them in code, and experimenting is exactly what we will be doing during the tour.

The tour is meant to be light, breezy, and reassuring. As I said, if I show you something that seems opaque or confusing, pass it by. Everything we see during the tour will be thoroughly covered in later chapters.

3.1.1   Results of expressions are displayed when not stored

If you type an expression but do not store the result, the result is displayed:

```
: 2+2
  4
: x = 2+2
: x
  4
```

Display of results when they are not stored happens in programs, too. The program `hello()` could have read

```
void hello()
{
        "hello, world"
}
```

You may think of expressions as consisting of operators, functions, and the like—such as 2+2, `sqrt(2)`, and "Mary" + "lamb"—but simple things like 2, x, and "hello, world" are expressions, too. When I typed x by itself in the output above, x was an expression, and the result of the expression was displayed because Mata displays unstored expressions. In the same way, the modified `hello()` program will display "hello, world" minus the quotes.

You will not often use naked expressions in programs because functions like `printf()` offer more control over how the results are shown. Naked expressions are useful in debugging, however. Say you have written the following program:

```
real scalar nfactorial_over_kfactorial(real scalar n,
                                        real scalar k)
{
        real scalar     result, i
        if (n<0 | n>1.0x+35 | n!=trunc(n)) return(.)
        if (k<0 | k>1.0x+35 | k!=trunc(k)) return(.)
        result = 1
        for (i=n; i>k; --i) result = result*i
        return(result)
}
```

I know we have not yet discussed Mata's programming language, but you need not understand the code to understand the point I want to make. To debug the program, you might temporarily modify the program to read

```
real scalar nfactorial_over_kfactorial(real scalar n,
                                        real scalar k)
{
        real scalar     result, i
"n!/k! begins; n and k are"
(n, k)
        if (n<0 | n>1.0x+35 | n!=trunc(n)) return(.)
        if (k<0 | k>1.0x+35 | k!=trunc(k)) return(.)
"making calculation"
        result = 1
        for (i=n; i>k; --i) result = result*i
"done, result is"
result
        return(result)
}
```

The lines I added appear in boldface, but on my terminal, they would appear the same as the rest of the lines. I boldface lines to make the additions easier for you to spot. I boldface lines in programs to call your attention to them.

Having modified the program, now when I execute it, I will see

```
: y = nfactorial_over_kfactorial(5,3)
  n!/k! begins; n and k are
        1    2
```

```
1  |  5    3  |
```

```
  making calculation
  done, result is
  20
```

3.1.2 Assignment

x = 2+2 stores the result of the expression in x. Variables can hold nonnumeric results, too:

```
: greeting = "hello, world"
```

```
: greeting
  hello, world
```

The equals sign is Mata's assignment operator.

The entire phrase x = 2+2 is considered an expression. There are two operators in it, = (assignment) and + (addition).

A single equals sign means assignment in Mata. In some languages, the equals sign is also used for testing equality, such as an if statement asking whether x equals 2 or greeting equals "hello, world". In Mata, equality tests are indicated by double equals signs:

```
if (x==2) ...
if (greeting=="hello, world") ...
```

3.1.3 Multiple assignment

You can store the result of an expression in multiple variables simultaneously:

```
: x = y = 2+2
```

```
: x
  4
```

```
: y
  4
```

In programming contexts, x = y = ⋯ is convenient for initializing variables at the outset of a loop:

```
. . .
xsum = ysum = zsum = 0
for (i=1; i<=n; i++) {
        xsum = xsum + x[i]
        ysum = ysum + y[i]
        zsum = zsum + z[i]
}
. . .
```

The `for` statement starts a loop in Mata. Understanding `for`'s syntax is not important for this tour; nonetheless, `for (i=1; i<=n; i++)` means to loop, starting with `i=1`, continuing while `i<=n`, and incrementing `i` by 1 at the end of the loop. The loop begins at the open curly brace and ends at the close curly brace.

In another program, you might need to store results of calculations and then take different actions depending on the results. In Mata, you can code

```
if ( avg = (zsum = xsum+ysum) / (zn = yn + xn) == 0 ) {
        . . .
}
else {
        . . .
}
```

Did you follow that? What is coded above is equivalent to coding

```
zsum = xsum + ysum
zn   = yn   + xn
avg  = zsum / zn
if (avg == 0) {
        . . .
}
else {
        . . .
}
```

I do not often code lines as dense as the one above because they are too difficult to read. I would, however, code

```
avg = (zsum = xsum + ysum) / (zn = yn + xn)
if (avg==0) . . .
```

I find the above two lines perfectly readable.

3.2   Real, complex, and string values

3.2.1   Real values

Obviously, Mata provides real values. We have been using them.

```
: x = 2+2
: x
  4
: sqrt(2)
  1.414213562
```

Mata's real values correspond to Stata's `double`. The values are the computer's double-precision representation of mathematically real numbers over the range of roughly $-8.988e{-}307$ to $8.988e{+}307$. The number closest to 0 without being 0 is approximately $1e{-}323$.

The smallest and largest integers that can be stored in a real without rounding are $\pm 9{,}007{,}199{,}254{,}740{,}992$.

Real values can also contain missing values, which are `.`, `.a`, `.b`, ..., `.z`, where

$$\text{all nonmissing values} < \texttt{.} < \texttt{.a} < \texttt{.b} < \cdots < \texttt{.z}$$

3.2.2   Complex values

Mata has complex values:

```
: z = (2+3i)*(5-2i)
: z
  16 + 11i
: exp(1i*pi())
  -1
```

Most programmers do not use Mata's complex-number capabilities. If you are an exception, see appendix B.

3.2.3   String values (ASCII, Unicode, and binary)

Mata has strings:

```
: s = "Mary had a lamb"
: s
  Mary had a lamb
: substr("Mary had a lamb", 6, 3)
  had
```

Mata's strings can be ASCII, Unicode, or binary. Strings may be 0 to 2 billion bytes long.

There is no special missing value for strings, although programmers sometimes use "", which is a string of length 0.

Mata has all the usual functions for processing ASCII strings, including substr(), strupper(), strlower(), etc. The functions listed extract substrings and convert strings to upper- and lowercase.

Mata also has Unicode string functions: usubstr(), ustrupper(), ustrlower(), etc.

And Mata has other string functions, such as functions for formatting output or performing file I/O.

In programming languages such as C and C++, processing binary data requires special programming techniques. In Mata, no special treatment is needed. Binary data can be stored in strings and manipulated with the standard string functions. For instance, substr() can extract a substring from a binary string just as it can from an ASCII string.

Mata's ability to handle binary strings in the same way as ordinary text strings is so surprising that it is worth demonstrating. Below, I interactively read the first 200 bytes from binary file auto.dta and display them.

```
: fh = fopen("auto.dta", "r")
: s = fread(fh, 200)
: fclose(fh)
: s
  <stata_dta><header><release>118</release><byteorder>LSF
> </byteorder><K>\uc\u0</K><N>J\u0\u0\u0\u0\u0\u0\u0</N>
> <label>\u14\u01978 Automobile Data</label>
> <timestamp>\u1121 Dec 2015 12:37</timestamp></header>
> <map>\u0\u0\u0\u0\u0\u0\u0\u0\xb2\u0\u0\u0\u0\u0\u0\u0-
```

The C programmers among you will be surprised by all the instances of \u0 appearing in the result. \u0 is how Mata displays binary 0. In C, binary 0 ends strings. In C, had we displayed this 200-byte string, the first binary 0 would have prematurely ended it. We would have seen

```
<stata_dta><header><release>118</release><byteorder>LSF
</byteorder><K>\uc
```

We did not see this because, to Mata, \u0 is just another character, no different from the character "a" or "b". In fact—and this will truly shock the C programmers—even strlen(s) knows that s is 200 bytes long:

```
: strlen(s)
  200
```

All of which is to say that Mata does not use binary 0 to mark the end of strings. That feature plus strings being allowed to be 2 billion bytes long plus `strlen()` knowing the length plus Mata's `bufio()` functions makes writing programs to process binary files easier.

Strings, whether ASCII, Unicode, or binary, may be added or multiplied:

```
: "abc" + "def"
  abcdef
: 2*"abc"
  abcabc
```

3.3   Scalars, vectors, and matrices

We have used scalar values in the examples so far. Mata also allows vectors and matrices. Vectors and matrices are entered using commas to separate the elements within rows and backslashes to separate the rows:

```
: x = (1,2,3)
: x
        1   2   3
    ┌─────────────┐
  1 │ 1   2   3   │
    └─────────────┘

: y = (1\2\3)
: y
        1
    ┌─────┐
  1 │ 1   │
  2 │ 2   │
  3 │ 3   │
    └─────┘

: X = (22, 2.9 \ 17, 3.35 \ 22, 2.64)
: X
         1        2
    ┌──────────────────┐
  1 │  22       2.9    │
  2 │  17       3.35   │
  3 │  22       2.64   │
    └──────────────────┘

: invsym(X'X)
[symmetric]
                    1                  2
    ┌───────────────────────────────────────────┐
  1 │   .0182372198                              │
  2 │  -.1225979159        .8617434448           │
    └───────────────────────────────────────────┘
```

Vectors and matrices can have up to 281 trillion rows and columns if your computer has sufficient memory. The memory requirement for real vectors and matrices is eight bytes per element.

3.3.1 Functions rows(), cols(), and length()

Functions `rows()` and `cols()` return the number of rows and columns of a vector or matrix:

```
: rows(X)
  3
: cols(X)
  2
```

These functions can be used with scalars, too. In a sense, all variables are matrices in Mata. Scalars are 1×1 matrices, and vectors are $1 \times c$ and $r \times 1$ matrices.

Functions `rows()` and `cols()` can be used to obtain the length of vectors, but it is better to use `length()`:

```
: length(x)
  3
```

It is better to use `length()` because it is too easy to code `rows()` when you mean `cols()` or `cols()` when you mean `rows()`. You are thinking that v is a row vector, and the next thing you know, you have typed `while (i<=rows(v))` in a program when you should have typed `while (i<=cols(v))`. You will avoid the problem altogether if you use `length()` and type `while (i<=length(v))`.

`length()` can be used with matrices, but it seldom is. It returns the total number of elements. You can obtain the memory consumed by real matrix X by coding `8*length(X)`.

3.3.2 Function I()

Function $I(n)$ returns the $n \times n$ identity matrix:

```
: I(3)
[symmetric]
        1   2   3

    1   1
    2   0   1
    3   0   0   1
```

3.3.3 Function J()

Function J(r, c, $value$) returns an $r \times c$ matrix with elements equal to $value$.

```
: J(2, 3, 0)
         1    2    3
      ┌─────────────────┐
   1  │   0    0    0   │
   2  │   0    0    0   │
      └─────────────────┘
```

The third argument, $value$, need not be a scalar. It can be a vector or a matrix:

```
: J(2, 1, (3,4))
         1    2
      ┌───────────┐
   1  │   3    4  │
   2  │   3    4  │
      └───────────┘
```

```
: J(2,2, (3,4\5,6))
         1    2    3    4
      ┌─────────────────────┐
   1  │   3    4    3    4  │
   2  │   5    6    5    6  │
   3  │   3    4    3    4  │
   4  │   5    6    5    6  │
      └─────────────────────┘
```

The third argument can be of any type, so J(r, c, $value$) can also be used to create complex, string, and other types of matrices.

3.3.4 Row-join and column-join operators

We have seen that commas and backslashes are used for entering matrices, such as,

```
: X = (86,13\13,22)
: X
[symmetric]
         1    2
      ┌───────────┐
   1  │  86       │
   2  │  13   22  │
      └───────────┘
```

Commas separate elements in a row, and backslashes separate the rows.

Comma and backslash are in fact operators in the same way that + and * are operators.

The comma operator is called column join.

The backslash operator is called row join.

If I typed a+b*c, you know that means a+(b*c). In mathematical jargon, * takes precedence over +. In computer jargon, * binds more tightly than +. However you say it, a+b*c means a+(b*c).

Comma takes precedence over backslash in the same way. The expression a\b,c means a\(b,c).

What does a\(b,c) mean? It means that a forms the first row or rows of the result and (b,c) are the remaining row or rows. For that to work out, a must have the same number of columns as (b,c). If not, the expression has a conformability error.

Consider a simpler expression, such as

```
: X = (86,13\13,22)
```

The outside parentheses are optional. I typed them because I think they make the expression more readable, but Mata does not require them. Because comma takes precedence over backslash, Mata executes the expression (86, 13 \ 13, 22) like this:

1. Join 86 and 13 to form the row vector (86, 13).

2. Join 13 and 22 to form the row vector (13, 22).

3. Stack the two row vectors to form a 2×2 matrix.

Here are some other ways you might use the comma and backslash operators. You can column-join row vectors:

```
: a = 1,2
: b = 3,4,5
: a,b
        1   2   3   4   5
    +---------------------+
  1 |   1   2   3   4   5 |
    +---------------------+

: a,b,a
        1   2   3   4   5   6   7
    +-----------------------------+
  1 |   1   2   3   4   5   1   2 |
    +-----------------------------+
```

You can row-join row vectors:

```
: a = 1,2
: b = 3,4
: a\b
        1    2
   1    1    2
   2    3    4
```

```
: a\b\a
        1    2
   1    1    2
   2    3    4
   3    1    2
```

```
: A = a\b
```

You can column-join matrices:

```
: A
        1    2
   1    1    2
   2    3    4
```

```
: A,A
        1    2    3    4
   1    1    2    1    2
   2    3    4    3    4
```

You can row-join matrices:

```
: A\A
        1    2
   1    1    2
   2    3    4
   3    1    2
   4    3    4
```

You can row-and-column-join matrices:

```
: A, I(2) \ I(2), A
        1   2   3   4

    1 │ 1   2   1   0
    2 │ 3   4   0   1
    3 │ 1   0   1   2
    4 │ 0   1   3   4
```

3.3.5   Null vectors and null matrices

A row vector is $1 \times c$.

A column vector is $r \times 1$.

A matrix is $r \times c$.

In Mata, r and c can be 0!

A matrix can be 0×0, or 0×5, or 9×0; a column vector can be 0×1; a row vector can be 1×0. When r or c is 0, the vector or matrix is called a null vector or matrix.

Mata's J(r, c, *value*) function can be used to create null vectors and matrices. Specify r as 0, c as 0, or both as 0. Because null vectors and matrices have no elements, J() does not use the value specified, but it does use the value's type. The value's type determines the type of the result:

J(0, 0, .) creates a **real** null matrix, as does J(0, 0, 3).

J(0, 0, C(.)) creates a **complex** null matrix, as does J(0, 0, 1+2i).

J(0, 0, "") creates a **string** null matrix, as does J(0, 0, "text").

Rather usefully, J(0, 0, x) creates a null matrix of the same type as x.

Null matrices of different types are not equal to each other:

```
: x = J(0, 0, .)
: s = J(0, 0, "")
: x==s
  0
```

Operator == tests equality. It returns 1 or 0, meaning true or false. x and s are not equal to each other because x is real and s is string.

Null vectors and matrices can be useful in programs to handle extreme cases. Let's consider linear regression. The logic for linear regression is as follows:

- Assume that y = X*b, approximately, which is to say, y = X*b + e, where e is the error in the approximation.

- X is an $r \times c$ matrix containing r observations on the c variables.

- y is an $r \times 1$ vector containing the corresponding r observations of the dependent variable.

- Then the value of b that minimizes the sum of the squared error e is `invsym(X'X)*X'y`.

Thus, the line b = `invsym(X'X)*X'y` might appear in a program we write. We might run the program on a particular matrix X and y and the program report b. We have not discussed how to program Mata yet, but we can imagine a program that contains the line. In fact, we can do better than imagine it. We can type the line interactively to see it work. We could type

```
: X = ...
: y = ...
: invsym(X'X)*X'y
```

If we filled in values for X and y, when we type the last line, we might see

```
: invsym(X'X)*X'y
              1

    1 |       1
    2 |       5
    3 |       7
```

and that would mean $y \approx 1x_1 + 5x_2 + 7x_3$.

If we were really writing a program to calculate `invsym(X'X)*X'y`, we would need to ensure that our program behaved gracefully when given extreme problems, such as data with no observations or a model with no variables. In most programming languages, we would have to include extra code in the program to deal with those cases. We do not in Mata.

Let's first make the calculation when there are zero observations, meaning that X is $0 \times c$ and y is 0×1:

```
: X = J(0, 3, .)
: y = J(0, 1, .)
: invsym(X'X)*X'y
              1

    1 |       0
    2 |       0
    3 |       0
```

The above result says $y = 0x_1 + 0x_2 + 0x_3$, which is an acceptable result for this extreme case. Thus, we will not have to write extra code to skip the calculation when `rows(X)` is 0.

Now let's consider the other special case of no variables:

```
. X = J(5000, 0, .)
: y = J(5000, 1, .)
: b = invsym(X'X)*X'y
: b
```

You cannot see it, but `b` is 1×0. Mata displays null vectors and matrices by displaying nothing. To convince yourself that Mata displayed nothing because the calculation produced a null result, you could type

```
: rows(b)
  0
: cols(b)
  1
```

The above is one feature of null vectors and matrices. We can ignore limiting cases, and usually, they will handle themselves adequately.

Null vectors are also useful when storing results accumulated in a loop. To show you this, I need to show you some Mata code, but the details of the code are unimportant. Say we have a loop that calculates a sequence z_1, z_2, ... that we need to store for later use. Assume that we do not know at the outset the number of terms we will need to store. We will accumulate terms until they are small enough. The code might read

```
zvals = J(1, 0, .)                    // zvals:  1 x 0
a = b = z = lastz = 0
while ( notconvergedyet(lastz, z) ) {
        olda = a ; oldb = b

        a = updateda(olda, oldb)
        b = updatedb(olda, oldb)

        lastz = z
        z = calcz(a, b)
        zvals = (zvals, z)            // add column to zvals
}
```

`zvals` starts as a 1×0 vector. The first time through the loop, the code column-joins `zvals` (1×0) with `z` (1×1) to produce a 1×1 result. `zvals` changes from being null to being 1×1. As the loop continues, `zvals` becomes 1×2, 1×3, and so on. When the loop concludes, `zvals` will be $1 \times$ `length(zvals)`. We will have `length(zvals)` values stored.

3.4   Mata's advanced features

Structures, classes, and pointers are Mata's advanced features. Calling the features "advanced" suggests they are unnecessary, yet many of the problems I write code for would be unprogrammable without them. I predict that you will become as dependent on these features as I am.

Before I can tell you about them, however, I need to explain Mata's variable types.

3.4.1   Variable types

A few of Mata's variable types are

>   real scalar
>
>   string vector
>
>   complex matrix

Variable types are composed of an element type and an organizational type, also known as eltype and orgtype. Any combination of eltypes and orgtypes is allowed, so the three examples above could just as well have been

>   string scalar
>
>   complex vector
>
>   real matrix

You will declare the variable types of arguments, variables, and returned results in the programs you write.

```
    function         first argument                  second argument
      returns              is                         is
          \                    \                   /
         real colvector bofX(real matrix X, real scalar i)
         {
    j and k are
                 \
                real scalar      j, k
                real colvector   b
              /
          b is



                        .
                        .
                        .

         }
```

`real colvector`, `real matrix`, and `real scalar` in the diagram above are declarations. When working interactively, declarations are not allowed.

When variable types are not explicitly declared, variables are given the default type, which is called

>   `transmorphic matrix`

`transmorphic` means the eltype can change:

```
: x = 4                    // x is currently real
: x = 1i                   // x is now complex
: x = "a"                  // x is now string
```

The `matrix` part of `transmorphic matrix` means the dimension can change, too:

```
: x = 4                            // x is currently real scalar
: x = (1i, 2i)                     // x is now a complex rowvector
: x = ("a", "b" \ "c", "d")       // x is now a string matrix
```

The overall variable type is any eltype and orgtype combination.

The eight eltypes are

Eltype	Comment
`transmorphic`	can morph into any of the types that follow
`numeric`	can morph between `real` and `complex`
`real`	
`complex`	
`string`	
`struct` *name*	
`class` *name*	
`pointer`	

The five orgtypes are

Orgtype	Dimension
`matrix`	$r \times c$
`vector`	$1 \times c$ or $r \times 1$
`rowvector`	$1 \times c$
`colvector`	$r \times 1$
`scalar`	1×1

r and c may be 0 or greater.

Orgtypes are less different from each other than eltypes. Eltype `real` is substantively different from eltype `string`. Meanwhile, there is just one substantive orgtype, `matrix`.

The other orgtypes are restrictions. For instance, a variable declared as a `real matrix` can pass as a vector or a scalar if the matrix happens to be 1×1.

3.4.2   Structures

Structures are variables that themselves contain other variables. Structures are defined by the programmer, such as `struct coord` in

```
struct coord
{
        real scalar x
        real scalar y
}
```

The definition creates a new eltype. This one creates a `struct coord`. Once defined, you can use the new eltype in programs you write:

```
real scalar foo()
{
        struct coord scalar    c
        real scalar            d
        .
        .
        .
        c.x = ...
        c.y = ...
        .
        d = sqrt(c.x^2 + c.y^2)
        .
        d = distance_from_origin(c)
        .
        .
        .
}
```

`c` is a `struct coord` scalar because we declared it that way. The declaration is required. Most structure variables are declared to be scalars, but vector and matrix are allowed.

Once `c` is declared, `c.x` and `c.y` are how you refer to `c`'s member variables. The `.x` and `.y` parts of the names are from the structure definition.

`c.x` and `c.y` are real scalars because they were declared that way in `coord`'s definition. You can use `c.x` and `c.y` just as you would use any real scalar variable. You can assign values to them:

```
c.x = ...
```

You can use them in expressions. You can pass them to arguments to functions:

```
d = sqrt(c.x^2 + c.y^2)
```

Just as `c.x` and `c.y` are variables, `c` itself is a variable. You can pass `c` in its entirety to functions:

```
d = distance_from_origin(c)
```

For this to work, the function must be expecting a **struct coord** scalar. The function's definition might be

```
real scalar distance_from_origin(struct coord scalar c)
{
        return( sqrt( c.x^2 + c.y^2) )
}
```

Something quite remarkable happened here—let's pause a moment to appreciate it. We passed `c` to a function, and the function received the entire `c`. `c` has two member variables, but it could have had scores or hundreds of members. The function can use any or all of them.

Just as remarkably, you can also write functions to return an entire `coord`:

```
struct coord scalar unitlength(struct coord scalar c)
{
        struct coord scalar    toret
        real scalar            distance

        distance = distance_from_origin(c)

        toret.c1 = c.c1 / distance
        toret.c2 = c.c2 / distance
        return(toret)
}
```

Being able to pass and return entire structures is why you will want to use them in programs. Let's consider another structure that contains more variables:

```
struct regression_results
{
        real vector      b            // coefficients
        real matrix      VCE          // variance matrix
        real scalar      N            // # of obs
        real scalar      ndf          // numerator dof
        real scalar      ddf          // denominator dof
        real scalar      MSS          // model sum of squares
        real scalar      RSS          // residual sum of squares
        real scalar      R_squared // R-squared
        real scalar      s2           // mean sq error
}
```

We can pass the entire set of **regression_results** to subroutines we write, and we can obtain the entire set of **regression_results** from other subroutines.

Have you ever heard the advice that well-written code contains lots of short subroutines? One problem in following that advice is passing variables to subroutines when you have lots of them. Instead of following the advice, you settle for a single, long program. Structures are the way around the argument-passing problem. If variable r is **struct regression_results** scalar, you can pass r, and the subroutine will have access to all of r's variables.

We will discuss structures in chapters 10 and 11.

3.4.3   Classes

Classes are generalizations of structures. For those who already know what they are, Mata's classes allow public and private members, inheritance, shadowing, virtual functions, and polymorphisms.

For the rest of us, in addition to having member variables, classes can have member functions, too.

```
class regression
{
    public:
            real vector       b          // coefficients
            real matrix       VCE        // variance matrix
            real scalar       N          // # of obs
            real scalar       ndf        // numerator dof
            real scalar       ddf        // denominator dof
            real scalar       R_squared  // R-squared
            real scalar       s2         // mean sq error
        private:
            real scalar       MSS        // model sum of squares
            real scalar       RSS        // residual sum of squares

    public:
            void              calc()     // <-- member function

    private:
            real matrix       XX()       // <-- member function
            real colvector    Xy()       // <-- member function
}
```

This class contains member variables b, VCE, ..., RSS, and it contains member functions calc(), XX(), and Xy(). Some members are public.

Public member variables are referred to in the same way as structure's member variables. You code c.b, c.VCE, ..., c.s2, where variable c is a **class regression** scalar. You use the same c. prefix to refer to public member functions. To call calc(), you code c.calc().

Private member variables and functions, on the other hand, can only be accessed by member functions. This means that calc(), XX(), and Xy() are the only functions allowed to access MSS and RSS, and they are the only functions allowed to call XX() and Xy(). The goal of privacy is to make the code that provides b, VCE, ..., s2, and calc() safely modifiable in the future. Because users of the class cannot use or call the private members, you can subsequently modify the code of the entire class—up to and including deleting and adding new private member functions—with the certain knowledge that no user of the class will be affected as long as the public members continue to fulfill their advertised purpose. Computer scientists call this encapsulation.

Classes also allow inheritance, but we will ignore that for now.

Defining a class requires declaring it and writing the code for its functions:

```
class regression
{
   public:
        real vector      b            // coefficients
        real matrix      VCE          // variance matrix
        real scalar      N            // # of obs
        real scalar      ndf          // numerator dof
        real scalar      ddf          // denominator dof
        real scalar      R_squared // R-squared
        real scalar      s2           // mean sq error
   private:
        real scalar      MSS          // model sum of squares
        real scalar      RSS          // residual sum of squares

   public:
        void             calc()

   private:
        real matrix      XX()
        real colvector   Xy()
}
void regression::calc(real colvector y, real matrix X)
{
        real matrix      XXinv

        XXinv = invsym(XX(X))
        b     = XXinv * Xy(X, y)
        RSS   = sum((y-X*b):^2)
        s2    = RSS / (rows(X)-cols(X))
        .
        .
}
real matrix regress::XX(real matrix X)
{
        return(X'X)
}
real colvector regress::Xy(real colvector y, real matrix X)
{
        return(X'y)
}
```

Notice that member function calc() uses members b, RSS, s2, XX(), and Xy() without a c. prefix. Inside class functions, members are directly exposed.

Outside of member functions, you must use the *varname.membername* way of specifying variables, just as you would with a structure:

```
void myreg(real colvector y, real matrix X)
{
        class regression scalar    r
        real scalar                i

        r.calc(y, X)
        printf("s2 = %11.0g\n", r.s2)
        "Coefficient Vector"
        for (i=1; i<=length(r.b); i++) {
                printf("x%g    %10.0g\n", i, r.b[i])
        }
        .
        .
        .
}
```

We will discuss classes in chapters 12 and 13. By the way, the linear-regression calculation formula I have been using, `invsym(X'X)*X'y`, has remarkably poor numerical properties. We will discuss that later, too.

3.4.4   Pointers

Pointers are scary. Ask anybody. The truth of the matter, however, is that pointers seem complicated because they are invariably used in complicated programs, and those programs are complicated for reasons having nothing to do with their use of pointers. Pointers themselves are neither scary nor complicated. Pointers themselves are easy to understand, and they solve problems that can be solved no other way.

Assume that you were writing a linear-regression routine using the linear-regression class described above. If you completed development of the class, you would quickly discover that it would be convenient to put a copy of the X matrix in the class, but you would hesitate to do that because of its potential size. If X contains 100,000 observations on 100 variables, its memory footprint would be 380 megabytes. Can you afford burning 380 megabytes just to make a copy for your convenience? Pointers provide you with a better alternative. With pointers, you can create a synonym for X and put that in the class. The synonym will be every bit as convenient as a copy and, being a pointer, will require an insignificant 8 extra bytes of memory.

That one example should be enough motivation to learn about pointers, and as I said, pointers are easy to understand anyway.

A pointer is a variable containing the memory address of another variable. If you code,

```
: p = &a
```

then p will be a pointer containing the address of a. The & prefix in front of a is Mata's address of operator. Here is an example:

```
: a = 2
: p = &a
: p
  0xf80a8b8
```

The ugly 0xf80a8b8 is the hexadecimal address where a is stored. We do not care that the address is 0xf80a8b8, but we do care that p contains &a.

&a means the address of a.

*p means the contents of the address stored in p. The * prefix is Mata's pointer-dereferencing operator. Yes, that is the same asterisk as Mata's multiplication operator, but if Mata can understand the difference between a*a and *p, so can you.

*p means the contents of the address stored in p or, put another way, the contents of a:

```
: *p
  2
```

Yes, indeed, *p is 2, just as a is 2, and they are equal because p == &a, meaning that they are the same 2.

To complete the demonstration that *p is a synonym for a, let's change the contents of *p and check that a simultaneously changes:

```
: *p = 4
: a
  4
```

You are almost an expert on pointers. There is only one more thing to know.

Pointers that do not point to anything contain NULL. That NULL is different from the null of null vectors and matrices. Capital NULL is the null memory address.

Right now, p still points to a, but I am about to change that:

```
: p
  0xf814108

: p = NULL

: p
  0x0

: *p
                    <istmt>:   3010   attempt to dereference
> NULL pointer
r(3010);
```

Now you are an expert on pointers.

Allow me to offer some guidelines about pointer jargon. Pointers contain a memory address. You may say, "p contains the address of a" or "p points to a". You may even say star-p in conversation. *p is a. Meanwhile, the & operator is called the address-of operator, not the and or ampersand operator.

I will show you how to use pointers to conserve memory in section 10.10, and I will show you how to use them to create a matrix with a ragged right edge in section 18.4.2.2.

3.5 Notes for programmers

3.5.1 How programmers use Mata's interactive mode

I said at the start of the tour that Mata's interactive mode is of no use in formal programming. It is, however, of great use to Mata programmers. Programmers use interactive mode to experiment with features and functions before using them in programs. It is one thing to read a description in a manual and another to see it for yourself. Experimenting reduces the chance that you will write code based on a misconception.

You enter Mata by typing `mata:` or `mata`. The colon changes how Mata behaves if you subsequently make a mistake. Enter by typing `mata:`, and Mata will complain and exit back to Stata if you make an error. Enter by typing `mata` without the colon, and Mata will complain but not exit.

For experimenting, enter Mata by typing `mata` without the colon.

```
. mata
─────────────────────────────────────────── mata (type end to exit) ───
: _
```

When you are done, type `end` to return to Stata. I am not going to do that right now because I want to use interactive mode to show you something. We used Mata's `invsym()` function earlier in this chapter to obtain regression coefficients:

```
: invsym(X'X)*X'y
  (output omitted )
```

You can read about `invsym()` by typing `help mata invsym()`. `invsym`(A), says the manual, returns a generalized inverse of real, symmetric, positive-semidefinite matrix A.

Let's try inverting a full-rank matrix:

```
: X = (86, 13 \ 13, 22)
: X
[symmetric]
        1    2

  1     86
  2     13   22

: Xinv = invsym(X)
: X*Xinv
[symmetric]
        1    2

  1     1
  2     0    1
```

`invsym()` will work with singular matrices, too. You should try it.

The manual says that the function is for use with symmetric matrices. You should wonder what would happen if you attempted to invert a nonsymmetric matrix. Will the program stop? Will Mata crash? Will Mata issue a warning and proceed? The way to find out is to run an experiment. Make a nonsymmetric matrix and invert it.

```
: X = (1, 2 \ 3, 5)
: Xinv = invsym(X)
: Xinv
[symmetric]
        1    2

  1     5
  2     -2   1

: X*Xinv
        1    2

  1     1    0
  2     5    -1
```

`invsym()` does not stop, and it does not issue a warning. It produces a wrong answer!

You have learned an important lesson. If A might not be symmetric, your program needs to check before using invsym(). Mata has an issymmetric() function, so it is easy to check.

There is a general lesson to be learned, too. Mata, unlike Stata, does not burn computer time protecting you from yourself. If you need to check assumptions, it is your responsibility to check them. invsym()'s incorrect answer is an example of that attitude.

This does not mean that you should mindlessly write code to check every assumption. If you need to calculate invsym(X'X), you know that X'X is necessarily symmetric. You include code to verify assumptions only when they might not be true.

3.5.2   What happens when code has errors

You just saw one possible outcome when you call functions incorrectly. They can return incorrect results. If you wrote code and used invsym() to invert a nonsymmetric matrix, your code would have gone on to produce its own incorrect result. The correct view of this case is that your code has an error, not invsym(). Your code used invsym() incorrectly.

Incorrect results are one possibility. Mata issuing an error message and aborting execution is the other. Here is an example of that:

```
: X = (1, 7 \ 2, 1 \ 9, -7)
: Xinv = invsym(X)
                 invsym():   3205   square matrix required
                 <istmt>:       -   function returned error
    r(3205);
```

I used invsym() to invert a rectangular (not square) matrix, which invsym() cannot do. The result was an error message.

You might be surprised that we did not see yet another incorrect result from invsym(). Why is that mistake handled differently from the previous one? The answer is that Mata's built-in functions do not go out of their way to produce incorrect results. They go out of their way to run quickly. Checking whether the matrix is symmetric is computationally expensive, and it is usually unnecessary because the matrices are known to be symmetric. Thus, it is the responsibility of the caller to check that the matrix issymmetric() in cases where it might not be. Checking dimensionality, meanwhile, is computationally cheap. invsym() and all of Mata's other functions check dimensionality as a matter of course.

The error output is called a traceback log. On the first line, the log shows where the error occurred and what it was:

```
    invsym():   3205   square matrix required
```

The second line shows why Mata was executing the function:

```
<istmt>:       -   function returned error
```

In this case, `invsym()` was executed because of the interactive statement I typed. If the error had occurred in a program, the traceback log might have been

```
invsym():  3205   square matrix required
       calc_b():      -   function returned error
   my_regression():      -   function returned error
```

This is the same error except that this time `my_regression()` called `calc_b()` and `calc_b()` called `invsym()`. Where is the error? Because `invsym()` is a Mata built-in function, the error is unlikely to be there. The offending code is more likely in `calc_b()` or `my_regression()`.

3.5.3   The _error() abort function

`_error()` is the Mata function that produces traceback logs and stops execution, a process known as abort with error.

In the previous section, we saw `invsym()` abort with error when the matrix to be inverted was not square. We will soon be writing our own functions and will have occasion to use `_error()` ourselves. We will use it by coding

```
if (situation irretrievable) _error(...)
```

We do not need to call `_error()` in most irretrievable situations because Mata aborts execution automatically. If we write a function that requires two arguments and the user specifies three, Mata aborts the function with error. If matrices A and B need to be conformable under addition and we write code that just proceeds to calculate A+B, Mata aborts if they are not conformable.

We need to call `_error()` when the function would not otherwise abort. If n simply has to be positive and we worry that the caller might specify a negative value, we code

```
      .
      .
      if (n < 0) _error(...)
      .
      .          .
```

`_error()` has three syntaxes:

```
_error(#)

_error("string")

_error(#, "string")
```

The first syntax displays the message associated with standard error code #, which is also known as a return code. Using this syntax requires finding the appropriate code and message. They are listed in **help m2 errors**.

The second syntax produces the custom message *string* that you specify. Code 3498 is displayed with the message.

The third syntax displays both the code and the message that you specify.

You can experiment with **_error()** interactively just as you can experiment with any function. Below, I experiment with variations for the argument being out of range:

```
. mata
─────────────────────────────────── mata (type end to exit) ───────
: _error(3300)
                  <istmt>:  3300  argument out of range
r(3300);

: _error("argument N out of range")
                  <istmt>:  3498  argument N out of range
r(3498);

: _error(3300, "argument N out of range")
                  <istmt>:  3300  argument N out of range
r(3300);
```

4 Mata's programming statements

4.1 The structure of Mata programs

Individual programs are formally called functions in Mata, but that will not stop us from calling them programs, routines, or subroutines. A program is a chunk of code. Here are some examples.

Function `speed_of_light()` takes no arguments and returns a value. It would be useful if you were an astrophysicist.

```
real scalar speed_of_light()
{
        return(299792458 /* m/sec */)
}
```

Function `show()` takes arguments but returns nothing. Functions returning nothing are common when displaying results or writing results to a file.

```
void show(real scalar a)
{
        printf("a = %f\n", a)
}
```

Function n_choose_k() and its subroutine nfactorial_over_kfactorial() really are
functions in the mathematical sense because they accept arguments and return results.

```
real scalar n_choose_k(real scalar n, real scalar k)
{
        return( n-k > k ?
                nfactorial_over_kfactorial(n, n-k) /
                nfactorial_over_kfactorial(k, 1)
                :
                nfactorial_over_kfactorial(n, k) /
                nfactorial_over_kfactorial(n-k, 1)
              )
}

real scalar nfactorial_over_kfactorial(real scalar n,
                                       real scalar k)
{
        real scalar     result, i
        if (n<0 | n>1.0x+35 | n!=trunc(n))  return(.)
        if (k<0 | k>1.0x+35 | k!=trunc(k))  return(.)
        result = 1
        for (i=n; i>k; --i) result = result*i
        return(result)
}
```

I want you to focus on the physical structure of the programs. That structure is

```
returnedtype  name(arguments)
{
        declarations
        program body
}
```

All Mata functions have this structure.

Most functions require *arguments* and return something. *returnedtype* specifies what is
returned, such as a **real scalar** or **complex matrix**.

Functions that return nothing are said to return **void**.

You can omit the *declarations* of the variables used in the body of the program, but
we will not omit them in this book. Omitting the declarations increases the chances of
mistakes, and programs without declarations sometimes run slower. They run slower
when the compiler—not knowing the type—needs to produce more general code that
can handle all the possibilities.

4.2   The program body

There are nine statements that can be used in the program body. They are as follows:

Conditional execution statements:

```
if (expr) ... else ...
```

Looping statements:

```
for (expr; expr; expr) ...
while (expr) ...
do ... while (expr)
continue              (continue with next iteration of loop)
break                 (break out of loop)
```

Go-to statements (useful when translating Fortran programs):

```
goto stmt
```

Exit and exit-and-return-value statements:

```
return   and   return(expr)
```

Assignment, subroutine calls, and the like:

```
expr
```

expr is an abbreviation for expressions.

4.2.1   Expressions

We will discuss expressions deeply in the next chapter, and anyway, you already know what expressions are. Examples of expressions include

```
i = i + 1
y = myfcn(x)
mysubroutine(a, b)
```

This last example may not look like an expression to you, but it is. It is an expression that returns `void`.

There is a lot I could tell you about expressions, but as I said, you mostly know what they are. I do need to tell you about three surprising features of expressions, however.

The first surprising feature is that mathematical expressions such as

```
(-b + sqrt(b^2 - 4*a*c)) / (2*a)
```

and logical expressions such as

```
a>1 & b<2
```

are, despite appearances, both numerical expressions. They are numerical because they both return numerical results. Logical expressions return 1 or 0, where 1 means true and 0 means false. Mathematical and logical expressions may differ in the operators used, but they do not differ in the type of results they produce. Because they do not differ, mathematical and logical operators can be combined in surprising and useful ways.

For instance, say you have three numerical variables, a, b, and c. How many are negative? Answer: (a<0) + (b<0) + (c<0) are negative.

Arithmetic expressions can substitute for logical expressions, too. A condition is deemed to be true if the expression evaluates to any value except 0 because 0 means false. This means you can code

```
if ( (-b + sqrt(b^2 - 4*a*c)) / (2*a) ) ...
```

and what follows the if will be executed when (-b + sqrt(b^2 - 4*a*c)) / (2*a) is not 0.

The equivalency of numeric and logical expressions is Mata's first surprising feature. The second is that = means assignment and == means equality. Do not code

```
if (x=2) ...
```

when you mean

```
if (x==2) ...
```

The first is not an error; it is a bug. Mata will not complain when you code if (x=2), but the code will not do what you expect. The code will treat x=2 as assignment, meaning x will be changed to be 2. If that is not bad enough, assignment leaves behind the value, so the expression will be treated as true.

Coding x==2 is how you ask whether x is equal to 2.

Coding x!=2 is how you ask whether x is not equal to 2.

Finally, I need to tell you about Mata's ++ and -- operators. Coding i++ increments i by 1. You can think of it as a shorthand for i = i + 1. By the same token, coding i-- decrements i by 1.

Later in this chapter, I will show you examples of i++, such as

```
for (i=0; i<=n; i++) ...
```

I could just as well present the example as

```
for (i=0; i<=n; i=i+1) ...
```

Most programmers type `i++` instead of `i = i + 1`.

You can code the `++` operator after the variable name or before it: `i++` or `++i`. When coded as a standalone statement, which you code makes no difference. Coded in the midst of an expression, there is a distinction. Look at the following two statements:

```
z = v[i++] + x

z = v[++i] + x
```

`v[i++]` means obtain `v[i]` and then increment `i`.

`v[++i]` means increment `i` and then obtain `v[i]`.

For instance, if `i` were 2 before the statements were executed, then

> `v[i++]` accesses `v[2]`, whereas
> `v[++i]` accesses `v[3]`,
> and either way, `i` is incremented to be 3.

`i--` and `--i` work the same way.

4.2.2 Conditional execution statement

The syntax of `if (`*expr*`)` ... `else` ... is

```
        if (expr) stmt1
and
        if (expr) stmt1
        else stmt2
```

stmt1 is executed if *expr* evaluates to true (nonzero).

When `else` is coded, *stmt2* is executed if *expr* evaluates to false (zero).

You can code

```
if (x==2) y = myfcn(z)
```

and you can code

```
if (x==2) y =  myfcn(z)
else      y = altfcn(z)
```

To specify multiple statements following the `if` or `else`, enclose them in braces:

```
if (x==2) {
        y1 = myfcn(z1)
        y2 = myfcn(z2)
}
else {
        y1 = altfcn(z1)
        y2 = altfcn(z2)
}
```

You can use braces even when there is only one statement. Rather surprisingly, braces can even contain no statements:

```
if (x == 2) {
}
else x = 1
```

The code would be more readable, however, if you simply coded

```
if (x != 2) x = 1
```

Mata users often code

```
if (x) ...
```

instead of coding

```
if (x!=0) ...
```

because expressions are considered true when not 0. Omitting the `!=0` is considered good style.

4.2.3   Looping statements

Mata has three looping constructs: `for`, `while`, and `do ... while`.

`for` and `while` are the most commonly used. They check the condition for repetition of the loop at the outset, meaning the loop may not be executed at all. This is usually an advantage because it handles extreme cases elegantly.

`do ... while` checks the condition for repetition at the end of the loop and is for those situations where you need to make a trip through the loop at least once.

4.2.3.1   while

The syntax of `while` is

        `while` (*expr*) *stmt*

where *stmt* is one statement or is multiple statements enclosed in braces,
just as with `if`.

`while` works like this:

1. Execute *expr*. If *expr* is false (0), go to step 4.

2. Execute *stmt*.

3. Go to step 1.

4. ....

Here is an example of `while`:

```
x = 1                              // set initial value
while (abs(f(x)) > 1e-8) {
        x = x + f(x)/fprime(x)
}
```

Iterative approximations are often programmed using `while`. This example uses the
Newton–Raphson method to find the value of `x` that makes `f(x)==0`. As coded, the
loop continues as long as $|f(x)| > 10^{-8}$, so the value of `x` found will not literally make
`f(x)==0`, but it will make `f(x)` be close to it.

The line `x = x + f(x)/fprime(x)` is the body of the loop. The line calls two functions,
`f(x)` and `fprime(x)`. `f(x)` returns `f()` at the current value of `x`, the function we are
seeking to make 0. `fprime(x)` returns its derivative.

The above loop could be used to solve for the numerical value of the square root of 2.
`f(x)` would be `x^2-2` and `fprime(x)` would be `2*x`. This problem is so simple that we
can dispense with the functions and simply substitute the expressions into the code:

```
x = 1                        // set initial value
while (abs(x^2-2) > 1e-8) {
        x = x + (x^2-2)/(2*x)
}
```

We can even try this interactively:

```
: x = 1
: while (abs(x^2-2) > 1e-8) x = x - (x^2-2)/(2*x)
: x
  1.414213562
```

Most programmers needing the square root of 2 would simply code `sqrt(2)`. I merely wanted to illustrate the `while` loop in an interesting way. `while` (*expr*) *stmt* repeatedly executes *stmt* as long as *expr* is nonzero.

When there is only one *stmt* following the `while` (*expr*), you can type it in braces or omit them. I showed both ways above. When *stmt* includes multiple statements, you must enclose them in braces.

It can happen that you will have no lines at all in the body of the loop. Consider the following example:

```
while (one_NR_step(x) > 1e-8)
```

Imagine that `one_NR_step(x)` updates x by taking a Newton–Raphson step and returns `abs(x^2-2)` evaluated at the updated x. Then the above code would also solve the square-root-of-2 problem. As coded, however, it will not work, because when the loop has no body, you must make that explicit. You can do that in either of two ways. One way is to explicitly include an empty body in braces:

```
while (one_NR_step(x) > 1e-8) { }
```

The other is to explicitly state "and the statement ends here" by coding a semicolon at the end of the `while` statement:

```
while (one_NR_step(x) > 1e-8) ;
```

Mata does not require semicolons at the end of statements, but you can code them when you wish, or when they are needed.

You must use one of these two approaches when the body of the loop is empty. If you do not, results can be surprising. A piece of your code might read

```
while (one_NR_step(x) > 1e-8)
next_programming_statement
another_programming_statement
```

As written, Mata will interpret the code as if it read

```
while (one_NR_step(x) > 1e-8) {
        next_programming_statement
}
another_programming_statement
```

That is, Mata will take *next_programming_statement* as being the body of the loop. To avoid that, code

```
while (one_NR_step(x) > 1e-8) { }
next_programming_statement
another_programming_statement
```

or code

```
while (one_NR_step(x) > 1e-8) ;
next_programming_statement
another_programming_statement
```

When you need to code an endless loop, code 1 for the expression:

```
while (1) {
        ...
}
```

No programmer wants to create truly endless loops, of course, but programmers code seemingly endless loops when the body contains the code that will cause the code to exit. Here is an example:

```
while (1) {
        s = get_next_line()
        if (s=="") return(error_no_end_stmt())
        if (s=="end") break
        if (xeqline(s)) return(error_line_invalid())
}
// break takes us here
```

There are three exits from this loop. The first and last are **return()** statements and cause the code to exit not just the loop, but the function as well. The middle exit is **break**, which exits just the loop, and which we will discuss later in this chapter.

4.2.3.2   for

The syntax of **for** is

---

for (*expr1*; *expr2*; *expr3*) *stmt*

where *stmt* is one statement or is multiple statements enclosed in braces.

expr1 and *expr3* may be empty, but *expr2* may not.

If *stmt* is empty, code empty braces or semicolon.

Consider the code

```
sum = 0
for (i=1; i<=length(v); i++) sum = sum + v[i]
```

Here is how you read the **for** statement aloud:

> For **i** equal to 1, while **i** is less than or equal to **length(v)**, execute
> **sum = sum + v[i]** and then increment **i**.

The role of each of the syntactical elements is

- *expr1*, which is **i=1**, specifies something to be done before the loop begins.

- *expr2*, which is **i<=length(v)**, specifies when the loop concludes but states the condition positively, as "the loop continues as long as *expr2* is true".

- The loop's body, which is **sum = sum + v[i]**, appears at the end. It can be one statement or it can be multiple statements enclosed in braces.

- *expr3*, which is **i++**, specifies something to be done at the conclusion of the body. In this case, that "something" is to increment **i**.

for is often used to iterate through a vector or matrix, as in

```
sum = 0
for (i=1; i<=length(v); i++) sum = sum + v[i]
```

If there is more than one statement in the body, it must be enclosed in braces:

```
sum = sum2 = 0
for (i=1; i<=length(v); i++) {
        sum  = sum  + v[i]
        sum2 = sum2 + v[i]^2
}
```

Note that the condition statement is executed at the top of the loop, and that means the body of the loop might not be executed at all. In this example, the body would not be executed if **length(v)==0**, which would happen if **v** were 0×1 or 1×0. In that case, the final result is that **sum** and **sum2** will be 0 just as they should be. This is an example of what I meant when I said that checking conditions at the top of loops handles extreme cases.

We previously used the Newton–Raphson method to iteratively solve for the square root of 2 using the **while** statement. We coded

```
x = 1
while (abs(x^2-2) > 1e-8) x = x - (x^2-2)/(2*x)
```

This logic could just as easily be coded using **for**. Here are three ways we could do that:

```
for (x=1; abs(x^2-2)>1e-8;) x = x - (x^2-2)/(2*x)
for (x=1; abs(x^2-2)>1e-8; x = x - (x^2-2)/(2*x)) { }
for (x=1; abs(x^2-2)>1e-8; x = x - (x^2-2)/(2*x)) ;
```

Of the three alternatives, I like the first one the best because I am not fond of loops with empty bodies. When the body has only one statement, as the first solution has, I sometimes code the entire loop in one line as shown above, and other times, I code it with braces, spreading the logic over three lines:

```
for (x=1; abs(x^2-2)>1e-8;) {
        x = x - (x^2-2)/(2*x)
}
```

My preferences aside, any of the styles is a valid and readable way to program the loop.

In the first solution, there is no *expr3*.

In the second solution, there is no body. What was the body was moved to *expr3*.

The third solution is the same as the second one except that the omitted body is indicated with a semicolon rather than with empty braces.

Whichever style you use, **for** produces the same results as **while**:

```
: for (x=1; abs(x^2-2)>1e-8;) x = x - (x^2-2)/(2*x)
: x
1.414213562
```

I showed you four examples of Mata programs at the beginning of this chapter. One of them used a **for** loop:

```
real scalar nfactorial_over_kfactorial(real scalar n,
                                       real scalar k)
{
        real scalar     result, i
        if (n<0 | n>1.0x+35 | n!=trunc(n)) return(.)
        if (k<0 | k>1.0x+35 | k!=trunc(k)) return(.)
        result = 1
        for (i=n; i>k; --i) result = result*i
        return(result)
}
```

The **for** loop in this example is interesting because the loop counts down, not up. The **for** loop reads

```
for (i=a-1; i>=b; --i) result = result*i
```

for's third expression is `--i`, which decrements `i`. If the loop had counted up, the third argument would have been `i++`. It makes no difference whether you code `i++` or `++i` in loops that increment and, in loops that decrement, whether you code `i--` or `--i`. I code `i++` and `--i`. I code `--i` to emphasize that the loop counts down.

4.2.3.3   do while

The syntax of do ... while is

```
do {
        stmt
} while (expr)
```

---

do ... while is a `while` loop with the checking of the continuation condition moved to the end of the loop. One trip is made through the loop, and thereafter, additional trips are made as long as *expr* is true.

Unlike `while` and `for`, the braces are required even if they enclose only a single statement.

You use do ... while when an outcome from a trip through the loop determines whether the loop is to be repeated. Here is an example:

```
do {
        get_and_execute(line)
} while(line != "exit")
```

Subroutine get_and_execute() fills in variable `line`, and the condition for repetition is that `line` is not equal to `exit`. This loop executes at least once.

In scientific programming, one often continues a loop until the error is small enough:

```
do {
        ...
        error = ...
} while (error > epsilon)
```

Here is another example where I instead continue until the error is not 0! The following code determines a computer's precision by iterating a calculation until numerical error arises:

```
a = 4/3
do {
        b = a - 1
        c = b + b + b
        eps = abs(c - 1)
} while (eps==0)
```

You could try the code interactively. On my computer, `eps` is 2.22045e–16 when the loop concludes. It will be that on your computer, too, because these days all computers use the same IEEE 754 standard. Back in the seventies, when the first version of this program was written, however, different computers used different standards, and it was sometimes important to write code that adapted itself to the computer's precision. This clever algorithm was in the original EISPACK (Garbow 1974) and written in Fortran. I translated the code to Mata.

4.2.3.4 continue and break

`continue` and `break` are used inside the bodies of `while`, `for`, and `do ... while` loops. Their syntaxes are

---

```
continue
break
```

`continue` specifies that the rest of the body of the loop is to be skipped, but the loop is to continue. Here is an example:

```
for (i=1; i<=length(X); i++) {
        j = mother_of(X, i)
        if (j == .) continue
        .
        .
        .
}
```

Vector X contains observations on related people. If the `i`th person's mother is not found in X, the loop proceeds to the next person because of the `continue` statement. Otherwise, the code does something useful regarding the child (`i`) and the mother (`j`).

`break` jumps out of the loop. The following code searches for the first `i` such that `v[i]==2`:

```
for (i=1; i<=length(v); i++) {
        if (v[i]==2) break
}
if (i > length(v)) {
        _error("v[i]==2 not found")
}
.
.
```

4.2.4   goto

Mata has a `goto` statement for one reason: so that you can translate old Fortran programs. Other than that, using `goto` is a recipe for producing bad code.

The syntax of `goto` is

---

goto *stmtname*

.

.

.

stmtname: *stmt*

where *stmtname* is the name of the go-to point, which can be in the code above or below the `goto` statement.

The go-to point is indicated by coding the *stmtname* followed by a colon.

The following Fortran code can be translated to Mata:

```
        A = 4.0e0/3.0e0
     10 B = A - 1.0e0
        C = B + B + B
        EPS = DABS(C - 1.0e0)
        if (EPS.EQ.0.0e0) GOTO 10
```

Making the translation is trivial. We translate the Fortran statement number 10 to Mata statement name s10, change Fortran's GOTO to Mata's goto, and make the other changes required because of the minor syntax differences between Fortran and Mata. The translation is

```
        A = 4.0e0/3.0e0
  s10:  B = A - 1.0e0
        C = B + B + B
        EPS = abs(C - 1.0e0)
        if (EPS == 0.0e0) goto s10
```

I showed you this code earlier in section 4.2.3.3, but you may not recognize it because I translated it more thoroughly:

```
a = 4/3
do {
        b = a - 1
        c = b + b + b
        eps = abs(c - 1)
} while (eps==0)
```

Code is easier to read if it is written in the modern style, yet I recommend that you do not make such translations because you can only introduce mistakes. Mata has a `goto` statement so that you do not have to translate to modern style.

4.2.5   return

The syntax of `return` is

---

  return(*expr*)

or

  return

Use `return(`*expr*`)` for functions that return a value.

Use `return` for functions that return nothing, that is, return void.

4.2.5.1   Functions returning values

Functions that return values must include at least one `return()` statement, such as

```
real scalar myfunction(...)
{
        real scalar result

        .
        .
        return(result)
}
```

Functions can include multiple `return()` statements:

```
real scalar myfunction(...)
{
        real scalar result
        if (x>=.)  return(.)
        if (x<0)   return(myfunction(-x/2))
        .
        .
        return(result)
}
```

`return()` stops execution of the function and returns the value specified to the caller.

4.2.5.2 Functions returning void

Functions that return nothing are said to return void. Here is an example:

```
void modify_in_place(real matrix A, real scalar k)
{
        real scalar    i, j
        for (i=1; i<=rows(A); i++) {
                for (j=1; j<=cols(A); j++) {
                        .
                        .
                        A[i,j] = ...
                }
        }
        return
}
```

The **return** statement in this function could be omitted because Mata assumes that void functions are to return at the end of the code. Omitting the end-of-function **return** statement is considered good style:

```
void modify_in_place(real matrix A, real scalar k)
{
        real scalar    i, j
        for (i=1; i<=rows(A); i++) {
                for (j=1; j<=cols(A); j++) {
                        .
                        .
                        A[i,j] = ...
                }
        }
}
```

The jargon for functions like the above is that they "fall off the end" of the routine.

Regardless of whether you code **return** at the end of the code, you can include other **return** statements to exit early:

```
void modify_in_place(real matrix A, real scalar k)
{
        real scalar i, j
        if (X[k,k] == 0) return

        for (i=1; i<=rows(A); i++) {
                for (j=1; j<=cols(A); j++) {
                        .
                        .
                        A[i,j] = ... A[i,j] ...
                }
        }
}
```

5   Mata's expressions

5.1   More surprises

Most lines in programs are expressions or at least contain them. Expressions can appear as statements by themselves, such as

```
i = i + 2

r = (-b + sqrt(b^2 - 4*a*c)) / (2*a)

y = myfcn(x)

mysubroutine(a, b)
```

and expressions appear in five of Mata's eight programming statements:

```
if (x>0) ...

for (i=1; i<20; i++) ...

while (abs(err)>1e-12) ...

do ... while (abs(err)>1e-12)

return(x/y)
```

All the above are straightforward and unsurprising examples.

I previewed a few surprising examples of expressions in the last chapter. Here they are again:

```
fraction = (++i)/n                                    (1)

a = ( b = sqrt(x^2+y^2) ) / sqrt(X^2+Y^2)             (2)

ratio = (numer=s1+s2) / (denom=n1+n2)                 (3)

pos = (a>0) + (b>0) + (c>0)                           (4)

if (i) ...                                            (5)

while (abs(err=x^2-2)>1e-12) ...                      (6)
```

Got it? Here is what each does:

1. `fraction = (++i)/n` is equivalent to

```
i = i+1
fraction = i/n
```

2. `a = ( b = sqrt(x^2+y^2) ) / sqrt(X^2+Y^2)` is equivalent to

```
b = sqrt(x^2+y^2)
a = b / sqrt(X^2+Y^2)
```

3. `ratio = (numer=s1+s2) / (denom=n1+n2)` is equivalent to

```
numer = s1 + s2
denom = n1 + n2
ratio = numer / denom
```

4. `pos = (a>0) + (b>0) + (c>0)` stores in `pos` the number of variables `a`, `b`, and `c` that are positive (or missing).

5. `if (i) ...` is equivalent to

```
if (i!=0) ...
```

It performs the body of the `if` only if `i` is not 0. For instance, the loop would be performed if `i` were -2, 0.5, 7, or missing value.

6. `while (abs(err=x^2-2)>1e-12) ...` is equivalent to

```
err = x^2-2
while (abs(err)>1e-12) {
        ...
        err = x^2-2
}
```

Here are some other expressions that might surprise you simply because they use unfamiliar characters. For instance, everyone knows that + means plus, but did you know that ! means not?

```
if (!(a==b & c==d)) ...                              (7)

if (!(sum(A:==0))) ...                               (8)

max = (a>b ? a : b)                                  (9)

divline = 55 * "-"                                   (10)

vec = 1..4                                           (11)

B = A´                                               (12)

X = A´A                                              (13)

Z = A#B                                              (14)
```

Explanations:

7. `if (!(a==b & c==d)) ...` executes the body of the `if` only if it is not true that `(a==b) & (c==d)`, which is to say, if `(a!=b) | (c!=d)`. `!` is Mata's logical negation (not) operator.

8. `if (!(sum(A:==0)))` ... executes the body of the `if` only if matrix `A` contains all nonzero elements. `A:==0` produces a matrix of 0s and 1s, where 1 denotes `A[i,j]==0` and 0 denotes `A[i,j]!=0`. `sum()` sums all the elements of the matrix, producing a scalar, which in this case equals the number of elements of `A` that are equal to 0. `!(...)` returns 1 only if `(...)==0`.

9. `max = (a>b ?  a :  b)` sets `max` equal to the maximum of `a` and `b`. If the logical expression before the `?` is true, then the first alternative is returned; otherwise, the second is returned.

10. `divline = 55 * "-"` sets divline to 55 dashes.

11. `vec = 1..4` stores in `vec` the row vector (1, 2, 3, 4). (`vec = 1::4` would store a column vector in `vec`.)

12. `B = A'` stores the transpose of `A` in `B`.

13. `X = A'A` stores `A'` multiplied by `A` in `X`. `A'A` has the same meaning as typing `A'*A`.

14. `Z = A#B` stores in `Z` the Kronecker product of matrices `A` and `B`.

5.2   Numeric and string literals

The word literal is computer jargon for values that appear in expressions, such as the 55 and the `"-"` in

```
divline = 55 * "-"
```

55 is a numeric literal and `"-"` is a string literal. In Mata, `55*"-"` produces 55 dashes. 55 is a literal because the symbols literally mean 55. `"-"` is a literal because the symbols literally mean a dash. Meanwhile, `x` in `x+1` is not a literal because `x` means the value stored in `x`.

5.2.1   Numeric literals

5.2.1.1   Base-10 notation

Here are some unsurprising examples of numeric literals:

2

3.14159

−7.2

2.213e+32

2.213E+32

```
1e–8

2.213d+32

2.213D+32

1d–8
```

The last six examples are written in E notation even though some of them are written using D rather than E. 2.213e+32 means 2.213×10^{32}. You can code the E in upper- or lowercase. 1e–8 means 1×10^{-8}. D and d mean the same as E. D dates back to the Fortran days of computing when D meant double precision and E meant single precision. In Mata, both mean double precision.

Numeric literals can also be any of Stata's missing values:

```
.
.a
.b
.
.
.
.z
```

All nonmissing values are less than the missing values. The statement $2 < .$ is true, for instance. For the missing values, $. < .a < .b < \cdots < .z$.

The following are valid statements:

```
a = 2

b = .

c = .a

if (a==2) ...

if (b==.) ...

if (b>=.) ...

if (b<.) ...
```

Note that b>=. asks whether b equals any missing value. b==. asks only whether b equals . in particular. b<. asks whether b is not missing.

5.2.1.2   Base-2 notation

Stata has a base-2 notation. I know this sounds arcane and like something you could safely skip. However, I am about to show you an easy way to calculate more-accurate results. Rather than code

```
d = (f(x+1.0e-8) - f(x)) / 1.0e-8
```

I am going to suggest that you code

```
d = (f(x+1.0x-1a) - f(x)) / 1.0x-1a
```

You do not yet know what `1.0x-1a` means, but certainly you can agree with me that substituting `1.0x-1a` for `1.0e-8` is easy to do. If you make that substitution, you will halve the numerical round-off error for reasonable values of x. For "unreasonable" values—say, x equal to 16 million—coding `1.0x-1a` will reduce the round-off error to 0.0000007 of the round-off error that using `1.0e-8` would produce. Said differently, if you use `1.0e-8`, you will have at most one digit of accuracy in the calculation; if you use `1.0x-1a`, you will have eight.

In what follows, I explain and justify the claims I just made.

Stata and Mata provide a base-2 notation called X notation. Many programmers, needing a small value, will code one of the following literals:

    1e–8

    1e–12

    1e–14

For instance, they might code

```
if (abs(x-y)<1e-14) ...
```

or

```
deriv = ( f(x+1e-8)-f(x) ) / 1e-8
```

The problem with 1e–8, 1e–12, and 1e–14 is that they cannot be represented precisely on a binary computer. The numbers stored are not 10^{-8}, 10^{-12}, and 10^{-14}; they are approximately $10^{-8}+5.17\times10^{-24}$, $10^{-12}+5.86\times10^{-28}$, and $10^{-14}+7.88\times10^{-30}$. Those small differences from the intended value can cause substantial numerical imprecision.

The issue with the decimal values 1e–8, 1e–12, and 1e–14 is that they have no exact representation in base 2. In base 2, they are repeating binaries in the same way that $1/7$ is a repeating decimal. $1/7$ in base 10 is $0.142857142857\ldots$

Numerical inaccuracy occurs because computers sometimes ignore the rightmost digits when performing addition and subtraction. You might do the same if I asked you to add 5.17 and 1.4231×10^{-4}. You might calculate

$$5.1700$$
$$+\ 0.0001[1231]$$
$$\overline{}$$
$$5.1701$$

If you were a base-10 computer with a five-digit accumulator, the above is precisely what you would do. You would ignore the digits I displayed in square brackets. Let me show how much subsequent numerical inaccuracy this can produce. Let's consider the calculation of $\{f(x+h) - f(x)\}/h$ when the function $f(\cdot)$ is $f(x) = x$. Mathematically, the calculation produces the value 1 for all values of x and h:

$$\{(x+h) - x\}/h = (x + h - x)/h$$
$$= h/h$$
$$= 1$$

Let's calculate $\{(x+h) - x\}/h$ for $x = 5.17$ and $h = 1.4231 \times 10^{-4}$ using five digits of accuracy. The result is

$$\{(x+h) - x\}/h = \{(5.17 + 0.0001[4231]) - 5.17\}/1.4231 \times 10^{-4}$$
$$= \{(5.17 + 0.0001) - 5.17\}/1.4231 \times 10^{-4}$$
$$= 0.7027$$

The correct answer is 1 and we just calculated 0.7027. How is it that we produced a result with not even its first digit correct when we performed the calculation with five digits of precision? We produced the result because we chose a poor value for h. Yes, h is just five digits—it is precisely 1.4231×10^{-4}—but those digits are all nonzero, and when we added $x + h$, we lost some of them. Meanwhile, we divided using all five of h's digits.

Had we chosen $h = 1.0000 \times 10^{-4}$, we would have obtained the correct answer of 1. We would have obtained the correct answer because when we added $x + h$ and ignored h's rightmost digits, those ignored digits would have been 0.

This leads to a rule that every scientific programmer should know:

> When choosing small values for use in numerical approximations, choose values with lots of trailing zeros.
>
> If you are using a base-10 computer, choose values 1.0×10^{-k}. No one uses a base-10 computer, however.
>
> If you are using a base-2 computer, choose values 1.0×2^{-k}.

The next time you are tempted to use numbers like 1×10^{-8}, 1×10^{-12}, and 1×10^{14}, you would be better off using 1×2^{-26}, 1×2^{-40}, and 1×2^{-46}, because those numbers are close to intended values and have a one-digit binary representation.

The X notation was added to Stata and Mata in Stata 6 (1999) for StataCorp internal use and was first documented in the Stata documentation in Stata 8 (2003). Mata was added to Stata in Stata 9 (2005). X notation is the equivalent of E notation but for base 2. X notation is actually a base-16 notation, but base 16 is equivalent to base 2 and numbers are shorter when written in base 16.

X notation	E notation
*mantissa*X$\pm$*exponent*	*mantissa*E$\pm$*exponent*
mantissa and *exponent* written in base 16	*mantissa* and *exponent* written in base 10
meaning:	meaning:
mantissa $\times$ 2^*exponent*	*mantissa* $\times$ 10^*exponent*

You can write numbers in X format, and you can display them in X format, too:

```
. display %21x 1e-13
+1.c25c268497682X-02c
```

I displayed 1e–13 using Stata, but I could just as well have displayed it from Mata:

```
: printf("%21x\n", 1e-13)
+1.c25c268497682X-02c
```

You do not need to be accustomed to working with numbers like 1.c25c268497682X–02c. I am not. On the other hand, I can tell at a glance that the closest one-digit binary number to 1.c25c268497682X–02c is 1.0x–2b. I will show you how to do this in a moment. The point I wish to make is this: if I were tempted to use 1e–13 in a program I was writing, I would instead use 1.0x–2b. I would do that for the same reason that I would use 1×10^{-4} if I were programming a base-10 computer with a five-digit accumulator, because I hope to obtain more-precise numerical results. They might even be whoppingly more precise, and even if they are not, they will be no less precise.

I will show you a real example, but first, let me show you how to find the closest one-digit binary to 1e–13:

1. Start with 1e–13 and display the result in %21x format. You will learn that 1e–13 is 1.c25c268497682X–02c.

2. Recall the meaning of 1.c25c268497682X–02c. It is

$$1.c25c268497682 \times 2^{-2c}$$

The base-16 digits are 0, 1, 2, ..., 9, a, b, c, d, e, and f. The period in 1.c25c268497682 is the base-16 point. In base 16, $8/16 = 0.8$ is one-half.

3. Thus, 1.c25c268497682 rounds to 2 because the part to the right of the base-16 point is greater than one-half. In base 16, $0.c > 0.8$.

4. And thus, the closest one-digit number to 1.c25c268497682 $\times 2^{-2c}$ is $2 \times 2^{-2c} = 2^{-2c+1} = 2^{-2b}$.

5. In X notation, 2^{-2b} is written as 1.0x–2b.

I said that you could do this at a glance, and you can once you strip away the explanation. The closest one-digit binary number to 1.c25c268497682X–02c is either 1.0x–2c or 1.0x–2b, depending on whether you round down or round up. Because 1.c25c268497682 > 1.8 (base 16), you round up. The closest is 1.0x–2b.

I suggest the following translations for 1e–8, 1e–12, and 1e–14:

Decimal value	Better value	Better value approximated in base 10
1.0e–8	1.0x–1a	1.490e–08
1.0e–12	1.0x–28	9.095e–13
1.0e–14	1.0x–2e	1.421e–14

I did not round the decimal values to better values as I prescribed because I know something about the intent of these decimal numbers. The popular small number 1e–8, for instance, is intended to represent the closest round decimal to half of double-precision accuracy. Half accuracy is precisely 1.0x–1a. 1.0e–14 is the closest round decimal that leaves two decimal digits of accuracy behind. 1.0x–2e does that. I have no idea what 1e–12 was originally intended to reproduce, so I rounded it in the same way we rounded 1e–13.

Using one-digit binary values improves accuracy in some cases and does not in others, but using them will never reduce accuracy. One case where using them improves accuracy is in calculating approximations to derivatives using

```
deriv = ( f(x+h) - f(x) ) / h
```

There are four sources of numerical round-off error in the above calculation.

1. There is error in the numerical calculation in x+h that we have been discussing. It will vary from 0 to a gargantuan 0.5, or even more, as I will show you.

2. There is error in calculating f(). Until we specify the function, we can only speculate about the error in its calculation. Without explanation, it usually varies from 1e–16 to 1e–12.

3. There is error in calculating the difference between f(x+h) and f(x). If h is chosen optimally, it will be 1e–8.

4. There is error in dividing by h, though it will be trivial. It will be 0 or 1e–16 in the worst case.

I am not going to present an analysis of all four sources of error here, but if you add up all the errors assuming the error from source 1 is 0, then deriv can be calculated to roughly eight digits of accuracy (worst case), meaning with a relative error of 1e–8. If you use decimal numbers like 1e–8 or 1e–12 for h, however, all bets are off.

Let me show you. In the table below, I set the error from sources 2–4 to 0, something that cannot be done in real code, so that we can focus on the error from source 1. The table reports a comparison of results with truth for h equal to 1.0e–8 and h equal to 1.0x–1a. 1.0x–1a is the binary-rounded value of 1e–8. The results are as follows:

	Relative errors in calculating deriv = (f(x+h)-f(x))/h due to error in calculating x+h	
	Relative error in calculated deriv given h	
	h	
x	1.0e–8	1.0x–1a
0	0	0
8	8.27e–08	0
64	6.27e–07	0
1,024	1.06e–05	0
32,768	2.83e–04	0
1,048,576	1.17e–03	0
4,194,304	2.45e–02	0
16,777,216	0.118	0
33,554,432	0.255	0
67,108,864	0.490	0
134,217,728	infinity	infinity

Notes:
1. Relative error = $|calculated - true|/true$, where *true* is the true value of the derivative.
2. Relative errors do not exist when *true* = 0. Absolute errors are 0 in that case.

The first thing to learn from this table is that relative errors are greater for h equal to 1.0e–8 than they are for h equal to 1.0x–1a. The relative error from using 1.0x–1a is 0 right up to the point that the calculation falls apart, which happens when x is so large that x==x+h.

Meanwhile, the relative error from using 1.0e–8 is 8.27e–08 when x is 8, a relative error equal to 8 times the total error we ought to see from real code! At 64, the error from using 1.0e–8 is 63 times the error we could achieve. At 1,024, it is 1,060 times. And so on. By 16,777,216, the relative error is 0.118, and we have crossed the line separating results that have round-off error from results that are simply wrong.

The lesson is that if you use small values such as 1.0e–8—values with no one-digit binary representation—no matter how much numerical analysis you know, no matter how many numerical techniques you use, and no matter how well you write the code, you will have undermined the calculation from the outset. It is true that there are formulas where using 1.0e–8 instead of 1.0x–1a makes no difference, but there is not a single formula in which using 1.0e–8 will produce a more accurate result on a binary computer than using 1.0x–1a will produce.

Do not use 1e–8, 1e–12, and 1e–16. Use 1.0x–1a, 1.0x–28, and 1.0x–2e.

It is beyond the scope of this book, but to achieve a guaranteed 1e–8 or smaller total error, h needs to be set optimally. The optimal value is the value such that $f(x + h)$ and $f(x)$ have roughly half their digits equal—their most significant ones—and differ in the remaining digits (Nash 1979, sec. 18.2).

5.2.2 Complex literals

Type i after a real numeric literal to make the literal imaginary:

```
2i
3.14159i
-7.2i
2.213e+32i
2.213E+32i
1.0x-2bi
1.0X-2bi
```

You add or subtract real and imaginary literals to form complex literals:

```
2 + 2i
2 - 7.2i
-1 + -7.2i
```

Missing values are allowed. Complex numbers contain a single missing value for the entire number, not one missing value for the real part and another for the imaginary part. The expression z>. returns true when z contains missing for both real and complex z.

When assigning real values to variables that are stored as complex, use Mata's make-complex C() function:

```
x = C(.)
x = C(.a)
x = C(2)
```

You need to do this because, in some cases, simply assigning a real value to x can result in x being recast to storage type real, and many of Mata's built-in operators and library functions return different results when given real or complex values. The square root of -2, for instance, is missing, whereas the same calculation on $-2 + 0i$ yields the appropriate imaginary result:

```
: (-2)^.5
  .
: (-2+0i)^.5
  1.41421356i
: (-2)^(.5+0i)
  1.41421356i
: sqrt(-2)
  .
: sqrt(-2+0i)
  1.41421356i
```

5.2.3 String literals

String literals are enclosed in quotes. "*text*" specifies a string literal equal to *text*.

Mata has two sets of quoting characters, called double quote and compound double quote. The standard double-quote character may be used to start and stop text when the text itself does not contain double quotes, as in

```
s = "Mary had a lamb"
```

You can use the double quotes to enclose strings, or compound double quotes:

```
s = `"Mary had a lamb"'
```

Open and close compound double quotes are each two characters:

Open quote	left single quote followed by double quote
Close quote	double quote followed by right single quote

Compound quotes may be used anywhere double quotes may be used. They must be used to specify strings that themselves contain double quotes or contain matched pairs of compound double quotes:

```
s = `"Mary had a lamb"´

s = `"Mary said, "I have a lamb""´

s = `"The first example was `"Mary had a lamb"´"´
```

If you need to create a string containing unbalanced compound double quotes, first create variables containing them:

```
openquote  = "`"   + `""""´
closequote = `""""´ + "´"
```

Then use the variables to construct the desired string, such as

```
s = "The left compound double quote is " + openquote
```

Regardless of the style of quotes you use to specify string literals, literals may specify an empty string:

```
s = ""
```

```
s = `""´
```

An empty string is truly empty. If you set

```
s = "" + ""
```

then s=="" would still be true.

5.3 Assignment operator

The equals sign is Mata's assignment operator:

```
a = 2

r = (-b + sqrt(b^2 - 4*a*c)) / (2*a)

z = exp(1i*pi())

s  = "Mary´s lamb"
```

Mata allows multiple assignments in expressions:

```
sumx = sumy = sumz = 0

avg = overallsum / (denom = n1 + n2 + n3)
```

The last expression is equivalent to coding

```
denom = n1 + n2 + n3
avg = overallsum / denom
```

Do not confuse assignment (=) with logical equality (==).

x = 2 assigns 2 to x.

x==2 asks whether x is 2 and returns 1 or 0.

5.4   Operator precedence

It is operator precedence that allows you to type a+b*c instead of a+(b*c). Multiplication has higher precedence than addition.

It is operator precedence that allows you to type r = a/b instead of r = (a/b). Division has higher precedence than assignment.

It is operator precedence that allows you to type a>2+3 instead of a>(2+3). Arithmetic operators have higher precedence than logical operators.

Operators, from highest to lowest precedence	
Subscripting operators	[]
Increment & decrement operators	--, ++
Structure & class element-access operators	->, .
Pointer operators	*, &
Arithmetic operators	', ^, - (negation), /, #, *, -, +
Logical operators	!, !=, >, <, >=, <=, ==, &, |
Row fill	..
Column join	,
Column fill	::
Row join	\
Ternary conditional	? :
Cast to void	(void)
Assignment	=

Notes:
1. See the sections below for operator precedence within category.
2. The colon operators are omitted from the table. Each colon operator has precedence just below the corresponding noncolon operator. For instance, :== follows == in logical operators, and :* follows * in arithmetic operators.

So how is *p[3]*2 interpreted? Look at the table above. Subscripting operators have the highest precedence, so *p[3]*2 means *(p[3])*2, meaning that p is a vector. Pointer operators have the next highest precedence, so *(p[3])*2 means (*(p[3]))*2.

And there you have it. The expression means to take the third element of vector p, treat it as a pointer, obtain what p[3] is pointing to, and then multiply that value by 2.

Do not code *p[3]*2 if you mean (*p)[3]*2, which would mean to treat p as a pointer scalar, take what it is pointing to, which must be a vector, take that vector's third element, and multiply that by 2.

Use parentheses when you are unsure about operator precedence. Even if you are sure, people who read your code might not be. Adding parentheses does no harm.

5.5   Arithmetic operators

Arithmetic operators, from highest to lowest precedence	
A'	transposition; conjugate transposition if A is complex
$a \hat{\ } b$	power
$-A$	negation
$A \ / \ b$	division
$A \ \# \ B$	Kronecker product
$A * B$	multiplication
$A - B$	subtraction
$A + B$	addition

Notes:
1. Uppercase letters designate scalars, vectors, or matrices. Lowercase letters designate scalars.
2. All operators may be used with real or complex.
3. Multiplication is listed above addition, and thus multiplication has higher precedence: A+B*C = A+(B*C).
4. Power takes precedence over negation: -a^b = -(a^b).

Capitalization in the table above and in tables to follow indicates where vectors and matrices are allowed. For instance, multiplication allows scalar, vector, or matrix operands because the operands are shown in uppercase: $A*B$. The power operator, on the other hand, is limited to scalars because the operands are shown in lowercase: $a\hat{\ }b$.

The usual matrix conformability rules apply when vectors and matrices are allowed. You cannot, for instance, add a 3×3 matrix and scalar:

```
 : I(3)+2
                <istmt>:  3200  conformability error
 r(3200);
```

The arithmetic operators can be used with real or complex values. Be aware that operators use the real interpretation when operands are real and use the complex interpretation when they are complex. With real operands, the square root of -1 is missing; with complex operands, it is 1i:

```
: (-1)^.5
  .
: (-1+0i)^.5
  1i
: (-1)^(.5+0i)
  1i
```

5.6   Increment and decrement operators

Increment and decrement operators, from highest to lowest precedence	
a++	use and then increment
a--	use and then decrement
++*a*	increment and then use
--*a*	decrement and then use

Note:
1. These operators may be used with real scalars only.

++ and -- are operators that increment and decrement variables.

i++ means i = i + 1; i is to be incremented.

i-- means i = i - 1; i is to be decremented.

i++ can be used on a line by itself:

```
sum = 0; i = 1
while (i <= n) {
        sum = sum + a[i]
        i++
}
```

In this case, i++ is equivalent to i = i + 1.

If you want to make the loop count down instead of up, you can code i-- or --i:

```
sum = 0;   i = n
while (i>=1) {
        sum = sum + a[i]
        --i
}
```

i++, ++i, i--, and --i can be used on lines by themselves, and they can be used inside expressions, such as

```
sum = 0 ; i = 1
while (i <= n) sum = sum + a[i++]
```

i++ means that i is to be incremented after it is used, so a[i++] means that the value a[i] is returned and then i is incremented.

a[++i] increments first and obtains the value of a[i] afterward. The loop above could be coded

```
sum = i = 0
while (i < n) sum = sum + a[++i]
```

a[i--] and a[--i] work the same way.

5.7 Logical operators

Logical operators, from highest to lowest precedence	
!A	logical negation; if A is scalar, true if A==0; if A is vector or matrix, then scalar operation performed on each element separately
A != B	true if A not equal to B
$a > b$	true if a greater than b
$a < b$	true if a less than b
$a <= b$	true if a less than or equal to b
$a >= b$	true if a greater than or equal to b
A == B	true if A equal to B
a & b	true if a!=0 and b!=0
$a \mid b$	true if a!=0 or b!=0

Notes:
1. Uppercase letters designate scalars, vectors, or matrices. Lowercase letters designate scalars.
2. When a logical expression is true, it evaluates to 1. When a logical expression is false, it evaluates to 0.
3. Operators may be applied to real, complex, or string.
4. The equality and inequality operators may be applied to any types, including pointer, structure, and class.
5. As a convenience to C programmers, || and && are synonyms for | and &.

The logical operators return 1 or 0, meaning true or false.

The equality and inequality operators may be applied to mismatched types and return that they are not equal; the operators will not produce error messages.

```
: "my string" == 2
  0

: mystructure == myclass
  0

: mystructure == 2
  0
```

When the equality and inequality operators are used to compare a real with a complex, the result is based on whether the two values are in fact equal, and not on how they are stored:

```
: (1+0i) == 1
  1
```

In addition to allowing mismatched types, the equality and inequality operators can be applied to nonconformable vectors and matrices:

```
: I(2) == (1, 0)
  0
```

Greater than, greater than or equal to, less than, and less than or equal to may not be used to compare across types:

```
: "mystring">2
type mismatch:   string > real not allowed
r(3000);
```

Greater-than and less-than operators may be applied to real, complex, and string scalars. When applied to complex, the complex absolute values—the lengths of the complex vectors—are compared.

```
: z1 = (1+1i)

: z1 < -z1
  0

: z1 > -z1
  0

: z1 == -z1
  0

: z1/2 < z1
  1
```

When greater-than and less-than operators are applied to strings, they are compared byte by byte. If the strings are of unequal length but otherwise equal, the shorter string is less than the longer string.

```
: "a">"b"
  0
: "alpha" > "alphb"
  0
: "alpha" > "alph"
  1
```

Mata deals with Unicode in the same way that Stata does. Mata has Unicode functions. If you are not familiar with them, see `help unicode`.

5.8   (Understand this ? skip : read) Ternary conditional operator

Ternary conditional operator
$(a \ ? \ B \ : \ C)$

Notes:
1. Uppercase letters designate scalars, vectors, or matrices. Lowercase letters designate scalars.
2. The expression evaluates to B if a!=0; otherwise, it evaluates to C.
3. If a!=0, C is not evaluated. If a==0, B is not evaluated. This means any assignments, increments, or decrements in the nonevaluated operand will not be performed.

I hope you are reading this if the title confused you, because I can only be sure that those who understood the title have already moved on.

The ternary conditional operator is

$(a \ ? \ B \ : \ C)$

An example is

```
(x<0 ? -x : x)
```

If the expression in front of the question mark is true (nonzero), then the first expression after the question mark is returned. Otherwise, the second expression is returned. The colon separates the alternative expressions.

Only one of the alternative expressions is executed. In the following example, `i` is incremented only if `i<N`:

```
(i<N ? a[i++] : 1)
```

Here are examples of ways that the ternary conditional operator is used:

```
absx = (x<0 ? -x : x)

sumpos = sumpos + (x[i]>0 ? x[i] : 0)

errprintf("%d %s not found\n", n, (n==1 ? "variable" :
            "variables"))

j = (side==1 ? m.j0 : m.j1-1)
```

5.9   Matrix row and column join and range operators

Matrix row and column join and range operators, from highest to lowest precedence	
$a \mathbin{..} b$	row fill
A , B	column join
$a :: b$	column fill
$A \setminus B$	row join

Note:
1. Uppercase letters designate scalars, vectors, or matrices. Lowercase letters designate scalars.

We discuss row and column fill after discussing row and column join.

5.9.1   Row and column join

Row and column join are used to build vectors and matrices.

Column join can be used to form row vectors:

```
rv1 = (1, 2, 3)

rv2 = (4, 5, 6)

rv3 = (rv1, rv2, 7, 8, 9)
```

Notice that **rv3** will be 1×9.

Row join can be used to form column vectors:

```
cv1 = (7 \ 8 \ 9)

cv2 = (10 \ 11 \ 12)

cv3 = (cv1 \ cv2 \ 13 \ 14 \ 15)
```

Row join can form matrices by row-joining row vectors. If we row join 1×3 vectors `rv1` and `rv2`, we form a 2×3 matrix:

```
mat1 = (rv1 \ rv2)
```

Column join can similarly form matrices by column-joining column vectors. If we column join 3×1 vectors `cv1` and `cv2`, we form a 3×2 matrix:

```
mat2 = (cv1 , cv2)
```

We can use both operators to form matrices directly:

```
mat3 = (rv1, rv2 \ 16, 17, 18, 19, 20, 21)
```

Note that comma has higher precedence than backslash, so the above is equivalent to

```
mat3 = (rv1, rv2) \ (16, 17, 18, 19, 20, 21)
```

In the examples above, the outside parentheses around the vectors and matrices are unnecessary. Nonetheless, it is better to type them because they sometimes are necessary and they always make code more readable.

5.9.2 Comma operator is overloaded

Comma has two meanings in Stata.

First, comma means the column-join operator that we have been discussing.

Second, comma is used to separate arguments in functions.

Assume that function `myeval()` takes two arguments, a matrix and a scalar. You might use the function by coding

```
B = myeval(A, 2)
```

In this case, comma separates the arguments. Assume you want to call `myeval()` with matrix $(1, 7 \setminus -4, 2)$. You could code

```
B = myeval((1,7\-4,2), 2)
```

In this case, different commas mean different things. The commas inside the inner parentheses mean column join, while the comma outside means argument separation.

If you omitted the inner parentheses,

```
B = myeval(1,7\-4,2, 2)
```

the statement would be an error because it would appear to Mata that `myeval()` was being called with four arguments, namely, 1, 7\-4, 2, and 2. Added spaces will help you see the line as Mata sees it:

```
B = myeval(1,   7\-4,   2,   2)
```

The code clearly states to send four arguments to `myeval()`, which is not the intent.

The parentheses around the matrix formation are necessary. There are other cases where the parentheses are necessary, too, so my recommendation is that you make it a habit of coding them in all cases.

5.9.3   Row and column count vectors

`..` and `::` are the row-fill and column-fill operators. They produce count vectors.

`1..4` produces row vector (1, 2, 3, 4). `4..1` produces (4, 3, 2, 1).

`1::4` and `4::1` produce the same results but as column vectors.

5.10   Colon operators for vectors and matrices

Colon operators, from highest to lowest precedence	
$A :\hat{\ } B$	power
$A :/ B$	division
$A :* B$	multiplication
$A :- B$	subtraction
$A :+ B$	addition
$A :!= B$	not equal
$A :> B$	greater than
$A :< B$	less than
$A :>= B$	greater than or equal to
$A :<= B$	less than or equal to
$A :== B$	equality
$A :\& B$	logical and
$A :\| B$	logical or

Note:
1.   Uppercase letters designate scalars, vectors, or matrices.

Colon operators perform the underlying operator element by element under relaxed conformability requirements.

Consider the colon-add (`:+`) operator:

1. If you colon-add a scalar to a matrix, the scalar is added to each element of the matrix.

2. If you colon-add a row vector to a matrix, the vector is added to each row of the matrix.

3. If you colon-add a column vector to a matrix, the vector is added to each column of the matrix.

4. If you colon-add a matrix and a matrix, the matrices are added element by element, which is the same result of simply adding the matrices.

The colon-multiply operator works the same way as colon-add. Operations are performed element by element, which means that when you colon-multiply two matrices, the result is not equivalent to matrix multiplication.

The logical colon operators can be surprisingly useful. How many elements of A are equal to 0? Answer: `sum(A:==0)` are equal to 0.

5.11   Vector and matrix subscripting

Subscripting operators, from highest to lowest precedence	
Element subscripting	
$A[r,c]$	(r,c) element of matrix A (r and c scalars)
$A[r,.]$	rth row of A
$A[.,c]$	cth column of A
$A[r,]$	synonym for $A[r,.]$
$A[,c]$	synonym for $A[.,c]$
$V[i]$	ith element of vector V
List subscripting	
$A[V,W]$	rows V and columns W of matrix A (V and W vectors or scalars)
$V[W]$	elements W of vector V (W vector)
Submatrix subscripting	
$A[\|\ X\ \|]$	submatrix of A from top left $(X[1,1], X[1,2])$ to bottom right $(X[2,1], X[2,2])$ (X a 2×2 matrix)
$V[\|\ W\ \|]$	elements $W[1]$ to $W[2]$ of V (W a 2×1 column vector)

Note:
1. Uppercase letters designate scalars, vectors, or matrices. Lowercase letters designate scalars unless otherwise noted.

The subscripting operators allow you to access elements of a matrix.

`A[1,2]` is how you access the $(1,2)$ element of matrix `A`.

`V[2]` is how you access the second element of vector `V`. You can also access the second element using `V[1,2]` if `V` is a row vector or using `V[2,1]` if it is a column vector.

You can even subscript scalars. Specifying a[1,1] or a[1] is equivalent to specifying a by itself. That is important because in a calculation producing matrix A, A could be a 1×1 matrix.

Subscripting is an operator in Mata, meaning that if invsym(X'X)*X'y produces a vector, then you can code (invsym(X'X)*X'y)[1] to access the first element of the vector.

Subscripted expressions may be used on the left or the right of assignment:

```
b = A[1,2] + 2
A[1,2] = b/c
```

Mata provides three ways of subscripting: element, list, and submatrix.

5.11.1 Element subscripting

Element subscripting operators, from highest to lowest precedence	
$A[r,c]$	(r,c) element of matrix A (r and c scalars)
$A[r,.]$	rth row of A
$A[.,c]$	cth column of A
$A[r,]$	synonym for $A[r,.]$
$A[,c]$	synonym for $A[.,c]$
$V[i]$	ith element of vector V

Note:
1. Uppercase letters designate scalars, vectors, or matrices. Lowercase letters designate scalars unless otherwise noted.

You can access single elements of matrix A by coding A[b,c] or, if A is a vector, A[b], where b and c are scalars. Note that vectors can be subscripted using one or two subscripts. If A is a row vector, you can code A[1,2] to access the second element, or code A[2].

Element subscripting is misnamed in that it can also produce entire rows or columns of a matrix when one of the subscripts is specified as missing value:

A[a,.]	refers to the ath row of A
A[.,b]	refers to the bth column of A

You can even access the entire matrix by coding A[.,.] although, with one important exception, you want to avoid doing that. More on the exception in a bit. Here is why you should avoid it. If you coded

```
B = A[.,.]*C
```

the result would be the same as coding

```
B = A*C
```

The result will be the same, but the first statement will take longer to execute and will temporarily use more memory. Hence, B = A*C is preferred.

Subscripted matrices can appear on the left of assignment, too, as in

```
B[.,.] = A*C
```

Coding the above might be the same as coding

```
B = A*C
```

but coding B[.,.] = A*C might also be an error.

When you code B[.,.] = A*C, you are specifying that the elements of existing matrix B be replaced with the elements of A*C. Copying elements not only takes longer than replacing the entire matrix, but if B does not have the same number of rows and columns as A*C, then Mata will stop executing your code and issue an error message.

When you code B = A*C, you are specifying that B in its entirety is to be replaced with A*C. It does not matter if B is 2×3 and A*C is 50×10. After the assignment, B will be 50×10.

Even so, there is one case (the exception I mentioned above) where coding B[.,.] on the left of assignment is absolutely necessary. That is when B is a view onto Stata's data. View matrices are created by the st_view() function:

```
st_view(A, ., (1,2))
```

The above code results in A being an $n \times 2$ matrix containing all the observations of variables 1 and 2 in Stata's memory. Or at least A will appear that way. Matrix A will actually be a view onto Stata's data and not a copy of what Stata has stored. There is another way that you could obtain a copy instead of a view, but views have advantages. Views consume less memory than copies, and changes to views change Stata's data. View matrices are one way that Mata programmers change the contents of Stata's data.

Let's imagine that the programmer wants to replace the values of the first two variables with new values stored in matrix B. This is a case when the programmer must code

```
A[.,.] = B
```

If the programmer coded A = B, then matrix A itself would be replaced by a new matrix. The values of the Stata variables would not be updated, and A itself would no longer be a view matrix. This is the one case when it is important that it is the elements of existing matrix A that are changed, and the programmer accomplishes that by coding A[.,.] = B. You can learn more about views in appendix A.

`A[.,.]` has only one use, but constructs such as `A[i,.]` and `A[.,j]` can be genuinely useful in all contexts. On the right side of assignments, they extract row `i` or column `j`. On the left, they provide an easy and fast way to update an entire row or column of an existing matrix.

Element subscripting is the fastest of Mata's subscripting methods.

5.11.2   List subscripting

List subscripting operators, from highest to lowest precedence	
$A[V,W]$	rows V and columns W of matrix A (V and W vectors or scalars)
$V[W]$	elements W of vector V (W vectors)

Note:
1. Uppercase letters designate scalars, vectors, or matrices. Lowercase letters designate scalars unless otherwise noted.

List subscripting is like element subscripting except that the indices themselves are allowed to be vectors. For instance,

```
: A
             1     2     3

    1        1     2     3
    2        4     5     6
    3        7     8     9
    4       10    11    12

: i = (1,4)                 // i is a vector
: j = (2,3)                 // j is a vector
: A[i, j]                   // select rows 1,4 of columns 2,3
             1     2

    1        2     3
    2       11    12
```

The `i` and `j` vectors used above are row vectors, but they could have been column vectors. Aesthetically, `i` should be a column vector and `j` should be a row vector, but Mata does not require that.

The index vectors can be defined inside the subscripts:

```
    : A[(1,4), (2,3)]
             1     2

    1        ?     3
    2       11    12
```

Remember to place parentheses around the index vectors. `A[1,4, 2,3]` would not be understood by Mata.

Missing value can be used to select all the rows or columns:

```
    : A[., (2,3)]
             1     2

    1        2     3
    2        5     6
    3        8     9
    4       11    12
```

```
    : A[(1,4), .]
             1     2     3

    1        1     2     3
    2       10    11    12
```

List subscripts can be used to index vectors, too:

```
    : v
             1     2     3     4

    1       10    20    30    40
```

```
    : v[(1,3,4)]               // select elements 1, 3, and 4
             1     2     3

    1       10    30    40
```

List subscripts can be used to duplicate rows or columns. I duplicate row 1 below:

```
    : A[1\4\1, .]
             1     2     3

    1        1     2     3
    2       10    11    12
    3        1     2     3
```

Finally, list subscripts can be used as the targets of an assignment:

```
: B
[symmetric]
            1     2     3     4

    1 |     1
    2 |     0     1
    3 |     0     0     1
    4 |     0     0     0     1

: C
            1     2

    1 |   100   101
    2 |   102   103

: B[2\3,  (2,3)]  = C
: B
            1     2     3     4

    1 |     1     0     0     0
    2 |     0   100   101     0
    3 |     0   102   103     0
    4 |     0     0     0     1
```

5.11.3   Permutation vectors

Permutation vectors are a programming technique used with list subscripts. They serve two purposes. They provide an efficient way to access data when the data are stored in a matrix, and they provide a more efficient way to implement permutation matrices in advanced mathematical programming.

5.11.3.1   Use to sort data

Consider a matrix containing rows and columns corresponding to observations and variables:

```
: Data
            1        2        3

    1 |    22     2930     4099
    2 |    17     3350     4749
    3 |    22     2640     3799
    4 |    20     3250     4816
    5 |    15     4080     7827
```

Column 1 is the first variable, column 2 is the second, and column 3 is the third. The observations are in a random order. It would be easy enough to put these data in order of increasing values of variables 1 and 2, by which I mean that observations are placed in ascending order of variable 1 and, within equal values of variable 1, placed in ascending order of variable 2. Mata has a function that will do that:

```
sort(Data, (1, 2))
```

sort(), however, is not the topic of this section. (Although, you can learn more about it by typing help mata sort().)

The topic of this section is permutation vectors. Permutation vectors are another way the observations could be reordered. A permutation vector is any vector that is a permutation on (1, 2, 3, 4, 5). For instance, consider this permutation:

```
: p
          1    2    3    4    5

    1  |  5    2    4    3    1  |
```

Vector p can be used with list subscripts to reorder the rows of Data. It would be more aesthetically pleasing if I had made p a column vector, but whether p is a row or column vector is of no importance to list subscripts, and so I opted to save paper.

In any case, we could reorder Data by typing

```
: Data[p,.]
             1        2        3

    1  |    15     4080     7827
    2  |    17     3350     4749
    3  |    20     3250     4816
    4  |    22     2640     3799
    5  |    22     2930     4099
```

Data[p,.] permuted the rows of Data. Had I instead typed, say, Data[., (2,1,3)], I would have permuted its columns. Had I typed Data[p, (2,1,3)], I would have permuted both its rows and its columns.

In textbooks on probability, permutations are often random. This permutation of Data, however, is anything but random. Look carefully, and you will see that Data[p,.] put the data in ascending values of its first two columns. I obtained the p to do this by having previously typed

```
p = order(Data, (1,2))
```

order() is a Mata function similar to sort() that, instead of returning the reordered matrix, returns the permutation vector that would reorder the matrix. You can learn more about order() by typing help mata order().

If I wanted to replace matrix `Data` with its permuted version, I could type

```
: Data = Data[p,.]
: Data
                1       2       3

    1          15    4080    7827
    2          17    3350    4749
    3          20    3250    4816
    4          22    2640    3799
    5          22    2930    4099
```

If all I wanted was sorted data, it would have been easier to type `Data = sort(Data, (1,2))`. However, I have another string matrix containing the names that go with `Data`:

```
: Names
                       1

    1         AMC Concord
    2           AMC Pacer
    3          AMC Spirit
    4       Buick Century
    5       Buick Electra
```

`Names` are no longer in the same order as `Data` because I reordered the rows of `Data`. I can reorder the rows of `Names` to match the reordered `Data` by using `p` to permute them:

```
: Names = Names[p,.]    // Names = Names[p] would have worked, too
: Names
                       1

    1       Buick Electra
    2           AMC Pacer
    3       Buick Century
    4          AMC Spirit
    5         AMC Concord
```

I could not have done that had I used `sort()`.

Reordering multiple matrices in the same way is one use of permutation vectors. The other is that they can be used to reestablish the original order because

the inverse of   `Data = Data[p,.]`

is   `Data[p,.] = Data`

Below, I restore both `Data` and `Names` to their original order:

```
: Data[p,.] = Data
: Names[p] = Names
: Data
              1        2        3

    1        22     2930     4099
    2        17     3350     4749
    3        22     2640     3799
    4        20     3250     4816
    5        15     4080     7827


: Names
                       1

    1          AMC Concord
    2            AMC Pacer
    3           AMC Spirit
    4        Buick Century
    5        Buick Electra
```

There is another way I could have restored the original order:

the inverse of	`Data = Data[p,.]`
is	`Data[p,.] = Data`
and is	`Data = Data[invorder(p),.]`

`invorder()` is the Mata function that calculates the inverse of a permutation vector; see `help mata invorder()`.

5.11.3.2 Use in advanced mathematical programming

Permutation vectors are sometimes used in mathematical programming. They are the computer programmers' equivalent of mathematicians' permutation matrices. Permutation matrices are row or column permutations of the identity matrix. If $\mathbf{P}$ is a permutation matrix, then $\mathbf{PX}$ permutes the rows of $\mathbf{X}$ and $\mathbf{XP}$ permutes the columns.

There is a one-to-one correspondence between programmers' permutation vectors and mathematicians' permutation matrices, and that is important for two reasons. Permutation vectors consume less memory than permutation matrices. A row-permutation matrix, for instance, is $r \times r$. The corresponding permutation vector has just r elements. In addition, reordering a matrix using a permutation matrix requires performing matrix multiplication, which requires substantially more computer time than reordering the matrix using list subscripts.

If you are programming a mathematical formula that uses permutation matrices, program it using permutation vectors.

Row-permutation matrices are formed by permuting $I(r)$. The equivalent permutation vector is formed by performing the same permutation on $1::r$ or $1..r$. Either will work because Mata does not care whether permutation vectors are columns or rows. Column-permutation matrices are formed by permuting $I(c)$. The equivalent permutation vector is formed by performing the same permutation on $1..c$ or $1::c$.

5.11.4 Submatrix subscripting

Submatrix subscripting operators, from highest to lowest precedence	
$A[\| \ X \ \|]$	submatrix of A from top left $(X[1,1], X[1,2])$ to bottom right $(X[2,1], X[2,2])$ (X a 2×2 matrix)
$V[\| \ W \ \|]$	elements $W[1]$ to $W[2]$ of V (W a 2×1 column vector)

Note:
1. Uppercase letters designate scalars, vectors, or matrices.

Submatrix subscripts are called range subscripts in the Mata documentation. Whatever you call them, they provide a fast way to extract a submatrix or to change a submatrix within a matrix.

$A[|i,j \ \backslash \ k,l|]$ specifies the submatrix from top-left corner (i,j) to bottom-right corner (k,l):

```
: A
              1      2      3      4

      1       1      2      3      4
      2       5      6      7      8
      3       9     10     11     12
      4      13     14     15     16


: A[|1,1 \ 3,2|]
              1      2

      1       1      2
      2       5      6
      3       9     10
```

The argument specified inside $[|$ and $|]$ is a 2×2 matrix. Comma and backslash are the ordinary column- and row-join operators. The subscript matrix can be stored separately:

```
: M
        1   2
    ┌─────────┐
1   │  1   1  │
2   │  3   2  │
    └─────────┘
```

```
: A[|M|]
        1   2
    ┌─────────┐
1   │  1   2  │
2   │  5   6  │
3   │  9  10  │
    └─────────┘
```

Vectors can be subscripted using 2×2 matrices in the same way, but it is easier to specify a column vector:

```
: v
        1    2    3    4    5
    ┌─────────────────────────┐
1   │  10   20   30   40   50 │
    └─────────────────────────┘
```

```
: v[|2\4|]
        1    2    3
    ┌───────────────┐
1   │  20   30   40 │
    └───────────────┘
```

Vector v is a row vector in the example. v[|2\4|] extracted v's second, third, and fourth elements. The same command expression, v[|2\4|], would have worked if v had been a column vector; in that case, a column vector would have been returned.

Submatrix subscripting can be used on the left of an assignment:

```
: A[|2,2\3,3|]  =  (600, 700 \ 1000, 1100)
```

```
: A
        1      2      3      4
    ┌───────────────────────────┐
1   │  1      2      3      4    │
2   │  5    600    700      8    │
3   │  9   1000   1100     12    │
4   │ 13     14     15     16    │
    └───────────────────────────┘
```

Submatrix subscripting can be performed using list subscripts. For instance, the last example could be accomplished by typing

```
: A[(2\3), (2,3)]  =  (600, 700 \ 1000, 1100)
```

Submatrix subscripting executes more quickly than list subscripting.

5.12   Pointer and address operators

Pointer operators, from highest to lowest precedence	
NULL	value of the NULL address
p->MBR	access structure or class member MBR
$(*p).MBR$	same as p->MBR
$*p$	access object at address p
$(*p)(\ldots)$	call function at address p, passing arguments $\ldots$
&$NAME$	obtain address of nonfunction object $NAME$
&$NAME()$	obtain address of function $NAME()$
&($expr$)	obtain address at which result of evaluating $expr$ is stored
&($NAME(\ldots)$)	obtain address at which result of running $NAME(\ldots)$ is stored (this line an implication of &($expr$))

Note:
1. Uppercase letters designate scalars, vectors, or matrices. Lowercase letters designate scalars.

Pointers were introduced in section 3.4.4. I said there that pointers were scary and then went on to convince you otherwise. This section, sadly, runs the risk of convincing you that you were right in the first place. There is a table below that is more friendly and will prove useful, especially later, after you have had some experience using pointers. Before I can present the table, however, we have i's to dot and t's to cross.

Pointer variables contain NULL or a valid address.

* is the pointer dereferencing operator for obtaining the contents once an address other than NULL is stored in the variable. -> is an alternative, more convenient way to dereference pointers that point to structures and classes.

& is the address-of operator that is used to obtain the addresses that are stored in pointer variables. & can be used to obtain the address of any Mata object, be it a variable, a function, or the result of running a function or evaluating an expression. There is an exception. You cannot obtain the address of a class member function, such as cl.foo(). You can obtain the addresses of class instances and of class instance member variables, but not of class member functions.

Use of the pointer operators is summarized below. Each statement includes the minimum number of parentheses required. Many users include extra parentheses for readability. Such a user might code, for instance, *(Q[3]) instead of *Q[3].

Statement	Meaning
1. Treatment of NULL	
p = NULL	set (reset) p to point to nothing
*p	*error:* cannot dereference NULL pointer
*NULL	*error:* cannot dereference NULL pointer
2. Treatment of scalars	
p = &a	address of variable a
*p	a
3. Treatment of vectors and matrices	
p = &B	address of vector B
*p	B
(*p)[3]	B[3]
p = &C	address of matrix C
*p	C
(*p)[2,3]	C[2,3]
p = &B[2]	address of a *copy* of B[2]
*p	contents of the *copy* made of B[2]
	note: *p is not a synonym for B[2]
4. Treatment of structures and classes	
p = &st	address of variable st (assume st is a structure or class)
*p	st
p->mbr	st.mbr
(*p).mbr	st.mbr (same as p->mbr)
p->sub.b	st.sub.b
(*p).sub.b	st.sub.b (same as p->sub.b)
p = &st.mbr	address of st.mbr
*p	st.mbr
5. Treatment of functions	
p = &foo()	address of function foo()
(*p)()	call foo() without arguments
(*p)(2, 3)	call foo(2, 3)
p = &cl.foo()	*error:* address of class function not allowed

(continued)

Statement	Meaning
6. Treatment of pointer vectors and matrices	
Q = J(1, 20, NULL)	1×20 pointer vector with each element set to NULL
Q[1] = &a	address of variable a
Q[2] = &st	address of variable st
Q[3] = &st.mbr	address of st.mbr
*Q	*error:* cannot dereference vector
*Q[1]	a
*Q[2]	st
*Q[3]	st.mbr
Q = J(5, 5, NULL)	5×5 pointer matrix with each element set to NULL
etc.	
7. Treatment of expressions	
p = &(2)	address of a 2
*p	2
p = &(2+3)	address of a 5
*p	5
p = &(I(20))	address of a 20×20 identity matrix
*p	20×20 identity matrix
*p[3,2]	$(3, 2)$ element of 20×20 identity matrix
p = &(*expr*)	address of result from evaluating *expr*
*p	result from previously evaluated *expr*
8. Treatment of pointers to pointers	
p = &a	address of variable a
q = &p	address of variable p
*q	address of variable a
**q	a

5.13   Cast-to-void operator

	Cast-to-void operator
(void) *expr*	act as if *expr* returns void

Coding

```
(void) somefunction(...)
```

calls **somefunction()** and discards the returned value if there is one. When functions return a value and it is not assigned to a variable, Mata displays what is returned. Placing (void) in front of the call prevents that.

6 Mata's variable types

6.1 Overview

Variable types appear in three places in programs, as indicated in the diagram below:

```
here                              here
    \                              /
   returnedtype  fcnname (arguments)
   {
              declarations
                        \
                    here

           program body
   }
```

For example,

```
real matrix fcnname(real matrix x, real scalar i)
{
        real matrix        result
        real scalar        j
        program body
}
```

Variable types are optional. The function could have been coded

```
function fcnname(A, i)
{
        program body
}
```

Serious programmers do not omit variable types. I admit that I sometimes omit them in do-files when I am writing short programs that I intend to use once and throw away, but I never omit them in serious work. Let me show you why.

Here is a program to determine whether `a*x^2 + b*x + c` has real roots:

```
function hasrealroots(a, b, c)
{
        return((b^2 - 4*a*c) > 0)
}
```

There is no error in the program if it is used the way I intended, which is for real values of a, b, and c. For instance, the program reports that $x^2 + 3x + 2$ has real roots:

```
: hasrealroots(1, 3, 2)
  1
```

The program reports that $x^2 + 3x + 4$ does not:

```
: hasrealroots(1, 3, 4)
  0
```

The problem with the program is that it returns incorrect results when a, b, or c is complex, and worse, the problem arises even when the imaginary parts are 0. The example above reports that `hasrealroots(1, 3, 4)` is 0, meaning that the corresponding quadratic has no real roots. If we specify `3+0i` instead of `3`, however, `hasrealroots()` reports 1:

```
: hasrealroots(1, 3+0i, 4)
  1
```

Both answers cannot be correct. (For your information, the first answer is the correct one.) We previously encountered calculations that produce different answers depending on how arguments are specified. In the previous chapter, I mentioned that `sqrt(-1)` evaluates to missing, which is the desired result on the real-number line, and that `sqrt(-1+0i)` evaluates to 1i, the desired result for the square root of -1 on the complex plane. You might wonder how Mata managed that when, mathematically, $-1 = -1+0i$. Mata managed it because typing -1 produces a number stored as real, whereas typing -1+0i produces the same number but stored as complex, and Mata makes calculations based on storage types.

Mata's based-on-storage-type logic produces a feature in the case of `sqrt()`. That same logic produces a bug in `hasrealroots()`. I did not notice the bug when I wrote the program because I intended to use the function with real values and tested it only with real values.

Let's trace how the wrong answer came about. If argument b is storage type complex, the entire calculation is performed on the complex plane. b^2 - 4*a*c was correctly calculated to be -7+0i. Then that result was compared with 0 by using the greater-than operator. The greater-than operator on the complex plane, however, is not the operator that you are familiar with. On the complex plane, the operator compares the "lengths" of the complex vectors. -7+0i has length 7, which is greater than the length of 0+0i, which is 0. Thus, hasrealroots() returned true.

And so it is that hasrealroots() has an arcane bug. The program was nonetheless fine for my use in a do-file. There is a difference, however, in writing programs for quick-and-dirty use by yourself and writing programs for use by others. Let's see what might happen if I made hasrealroots() available to other users. Assume that I give hasrealroots() to you, and I even warn that it is for use with real values only. Assume that you have values stored in Mata variables mya, myb, and myc. Taking my warning to heart, you examine the values to make sure they are real:

```
: (mya, myb, myc)
       1   2   3

  1  | 1   3   4 |
```

Satisfied, you use hasrealroots() to determine if there is a real-root solution:

```
: hasrealroots(mya, myb, myc)
  1
```

There is a real-root solution, says the program, but the program is wrong. What you failed to notice was the omitted "+0i" on the end of myb. I suppose you cannot be blamed for that, because Mata did not show it to you. Variable myb is nonetheless complex in the storage-type sense of the word. Allow me to show you:

```
: eltype(myb)
  complex
```

Now what would be your response were I to claim that the mistaken answer was your fault? Your response should be that the program I wrote is sloppy, and its sloppiness could result in real damage. Now put yourself in my shoes. What did I do wrong? Who would have guessed that -7 is greater than 0 on the complex plane? And in my defense, I did test the program thoroughly with real values.

What I did wrong was not restricting the program to run only with values that I knew it worked with. Restricting the program to run only with real values would have been easy. I just needed to declare the argument types:

```
real scalar hasrealroots(real scalar a, real scalar b,
                         real scalar c)
{
        return((b^2 - 4*a*c) > 0)
}
```

Had I declared the types, the result of you typing `hasrealroots(mya, myb, myc)` would have been

```
: hasrealroots(mya, myb, myc)
          hasrealroots():  3253  myb[1,1] found where real
>                                required
                <istmt>:      -  function returned error
r(3253);
```

Serious programmers use variable types to prevent programs from working in situations in which the programs are not intended to work or not known to work.

Another reason that serious programmers specify variable types is that the code will be more understandable if they come back to it later.

Yet another reason that serious programmers include variable types is that the compiler might tell them about errors in their code that they otherwise would not notice. In fact, serious programmers tell the compiler to *require* that variable types be specified. Here is an example:

```
: mata set matastrict on          // require variable types
: real scalar longandcomplicated(real matrix A, real scalar p)
> {
>           real scalar    i, j, r, c
>           real scalar    p1, p2
>           real matrix    R
>
>           r = rows(A)
>           c = cols(A)
(200 lines not shown)
>           return(R)
> }
variable p1 undeclared
r(3000);
```

In the above output, Mata reports that variable `p1` is undeclared. The bug, it turns out, is not that I forgot to declare `p1`, despite that being what the error message says. The bug is that I coded `pl` (the letter p and the letter l) when I meant to code `p1` (the letter p and the number 1).

Let's imagine that I had not instructed Mata to require variable types by setting `matastrict` on, and so the program compiles without mention of `pl`. It is easy to imagine scenarios like the following: Variable `pl` was used late in the code where `p1` needed to be reinitialized to 1. Because `p1` was not reinitialized, a loop was skipped. Skipping the loop produced incorrect results, but only in certain rare cases.

Without the warning message, I doubt I would have noticed my typographical error that was buried somewhere in 200 lines of code.

There are two things you must do if you want Mata to spot such problems in your programs. First, you must use explicit declarations. That is the topic of this chapter. Second, you must `set matastrict on`, which among other things, does away with allowing declarations to be omitted. I will tell you more about `matastrict` in the next chapter.

There is yet another reason to specify variable types, although it is the least important of the reasons. Specifying variable types allows Mata to produce more-efficient code in some cases.

6.2   The forty variable types

Mata has 40 variable types, which you construct by choosing one entry from column A and another from column B:

Column A: eltype	Column B: orgtype
`transmorphic`	`matrix`
`numeric`	`vector`
`real`	`rowvector`
`complex`	`colvector`
`string`	`scalar`
`pointer`	
`struct` *name*	
`class` *name*	

Some examples are `transmorphic matrix`, `real scalar`, `complex matrix`, `numeric vector`, `string colvector`, `pointer vector`, `struct mystruct scalar`, and `class myclass scalar`.

The term from column A specifies the type of the element, which is also known as the eltype.

The term from column B specifies how the element is organized, which is also known as the orgtype.

The five orgtypes are

Orgtype	Meaning
matrix	$r \times c$
vector	$r \times 1$ or $1 \times c$
rowvector	$1 \times c$
colvector	$r \times 1$
scalar	1×1

In the above, $r, c \geq 0$. If r or c is 0, then the matrix, vector, etc., is called a null matrix, null vector, etc.

All the orgtypes except matrix are special cases of other orgtypes. scalar, for instance, is a special case of colvector, rowvector, and vector, which are in turn each a special case of matrix. If a matrix is 1×1, it can pass for a scalar. If a function requires a scalar argument, you could pass a 1×1 matrix to it. By the same token, if a function requires a matrix argument, you could pass a scalar to it. The orgtypes are restrictions, and you specify the restriction that most narrowly describes the variable.

The eltypes are

Eltype	Meaning
real	real number
complex	complex number
string	string (ASCII, Unicode, binary)
pointer	pointer to (address of) another variable
struct *name*	structure object
class *name*	class object
numeric	real or complex
transmorphic	any of the above

The first six eltypes are mutually exclusive. A real and a string are obviously different from each other. Less obvious is that a real is different from a complex—and vice versa—even if both contain the same mathematically real value. That means you cannot pass a real to a function requiring a complex. If X is a real and you want to pass it to a function that requires a complex, you must specify C(X). C() is Mata's function to promote values to complex if they are not already complex.

The last two eltypes are more relaxed restrictions.

Eltype **numeric** means **real** or **complex**. A function requiring a **numeric** argument **X** will accept **X** whether it is **real** or **complex**. In this case, callers do not have to code **C(X)**. Instead, the programmer who writes the function allowing **numeric** arguments must use **C(X)** where appropriate in the code.

Eltype **transmorphic** amounts to no restriction at all. Any eltype will do. A variable declared **transmorphic** could be **real** at one point in the code, **complex** later, and **string** even later. Doing that in code is not the intent of **transmorphic**, however. Such use would only increase the chances of bugs. We will see appropriate use of **transmorphic** later in this chapter and then again in chapter 8.

6.2.1 Default initialization

Consider the following function:

```
real scalar foo()
{
        real scalar retval
        return(retval)
}
```

This function has a bug: variable **retval** is uninitialized. If you were to run **foo()**, you would discover that it returns missing value (.). Declared variables are pre-initialized by Mata with default values.

Mata initializes **scalar**s to contain the appropriate type of missing value, which is missing (.) in the case of **real** and **complex**, "" in the case of **string**, and NULL in the case of **pointer**.

Vector and matrix variables—**vector**, **rowvector**, **colvector**, and **matrix**—are initialized to be null, meaning they are dimensioned 1×0, 0×1, or 0×0.

Default variable initialization

Eltype and orgtype	Initialized value
`real scalar`	variable contains .
`complex scalar`	variable contains .
`numeric scalar`	variable contains .
`transmorphic scalar`	variable contains .
`string scalar`	variable contains ""
`pointer scalar`	variable contains NULL
`... vector`	variable is 1×0
`... rowvector`	variable is 1×0
`... colvector`	variable is 0×1
`... matrix`	variable is 0×0
`struct ... scalar`	member variables initialized as above
`class ... scalar`	member variables initialized as above

Despite the initialization outlined in the above table, we will use the word *uninitialized* to describe variables that programs do not explicitly initialize. The table above would be better called the uninitialized-value table, because Mata does not initialize variables to help you; it initializes them to protect itself from you. Consider a truly uninitialized `string scalar pointer s`. In C, if you copied a string into `s`, your program would crash. Do the same in Mata and you get a traceback log:

```
: mysystem()
        set_defaults():  3010  attempt to dereference NULL
>                              pointer
            setup():      -   function returned error
        mysystem():       -   function returned error
        <istmt>:          -   function returned error
r(3010);
```

The traceback log is the Mata equivalent of C code crashing. The only reason that Mata did not crash is because it pre-initialized the pointer to contain NULL.

6.2.2 Default eltype, orgtype, and therefore, variable type

Eltype defaults to `transmorphic` when not specified.

Orgtype defaults to `matrix` when not specified.

Therefore, variables default to `transmorphic matrix` when not explicitly declared.

6.2.3 Partial types

Because eltype defaults to `transmorphic` and orgtype defaults to `matrix`, specifying just an eltype and letting orgtype default to `matrix` or specifying just an orgtype and letting eltype default to `transmorphic` is allowed. As a result, the following are valid explicit declarations:

`transmorphic`	to mean	`transmorphic matrix`
`real`	to mean	`real matrix`
`colvector`	to mean	`transmorphic colvector`
`matrix`	to mean	`transmorphic matrix`

Not declaring variables is bad style. Is declaring variables by specifying just the eltype or just the orgtype good style?

The intent of good style is to reduce the chances of problems in your code. Implicit declaration is bad because there can be unintentional consequences. In section 6.1, `hasrealroots()` produced a wrong answer because it did not explicitly declare the eltype. That result argues that eltypes should be explicitly specified.

How about omitting the orgtype? I could show you programs that produce incorrect results when fed the unexpected vector or matrix, but such examples are rare. Even so, I do not recommend omitting it.

I do, however, think that `transmorphic` by itself is acceptable style. `transmorphic` specifies that no restriction be placed on the eltype and thus correctly suggests that no restriction is being placed on the orgtype either.

6.2.4 A forty-first type for returned values from functions

Variable types are also used to specify the type of value that functions return:

> *returnedtype fcnname* (*arguments*)
> {
>   *declarations*
>   *program body*
> }

A function can return any of the 40 variable types, including `transmorphic matrix`. Functions that return `transmorphic` do not have to return a variable that is declared `transmorphic`. They might return `real scalars` in some cases or `string vectors` in others.

Functions can also return `void`.

> `void` *fcnname* (*arguments*)
> {
>   *declarations*
>   *program body*
> }

`void` specifies that the function returns nothing. `printf()` is Mata's function to display formatted text:

```
: printf("F(%f, %f) = %f\n", ndf, ddf, F)
F(2, 48) = 3.15
```

Function `printf()` is `void`.

`fclose()` is Mata's function to close an open file. It too is `void`.

Mata can return values in the caller's arguments, a topic we will discuss in detail in section 8.2. Functions that do this often return `void`. `svd(A, U, s, Vt)` is Mata's function to calculate the singular-value decomposition of matrix `A`. It returns its results in `U`, `s`, and `Vt`. `svd()` is itself `void`.

`transmorphic` includes `void`, so `transmorphic` functions can return `void` in some cases and a result in others:

```
transmorphic foo(...)
{
        .
        .
        if (...) return       // return void
        .
        .
        return(x)
}
```

Functions that do this are usually considered bad style, but there are exceptions. Mata's `st_local()` function deals with Stata macros. When two arguments are specified, the function returns nothing and sets the contents of the Stata macro:

```
st_local(macroname, contents)
```

When one argument is specified, the function returns the contents of the Stata macro:

```
contents = st_local(macroname)
```

It is an elegant design in that the number of arguments determines whether the function returns anything, and so it is obvious to the caller when the function will return something and when it will return nothing. You could write your own functions following this design. To handle the return value, declare the function `transmorphic`; in the body of the function, code `return` where you want to return `void` and code `return(...)` everywhere else. We will discuss how to write functions with a varying number of arguments in section 8.3.

A function always returns nothing when it is declared `void`:

```
void myfcn(...)
{
        ...
}
```

It is not, however, an error if the caller of the function codes

```
x = myfcn(...)
```

`x` will be set to a 0×0 `real` `matrix`.

6.3   Appropriate use of transmorphic

When you do not explicitly specify function and variable types, they are assumed to be `transmorphic`. You can also explicitly declare them to be `transmorphic`. There are four valid reasons to declare them `transmorphic`:

1. Functions are declared `transmorphic` when they sometimes return one thing and other times return another. In the previous section, function `st_local()` returned the contents of a macro when one argument was specified, and set the macro and returned `void` when two arguments were specified.

2. Arguments of functions are declared `transmorphic` when they are allowed to be of one type or another. Such functions are said to be overloaded.

3. Arguments of functions are declared `transmorphic` when the arguments are used only for returning calculated results.

4. Arguments of functions are declared `transmorphic` when the variables contain values that will merely be passed along to other subroutines. Such variables are called passthru variables.

We discussed reason 1 in the section above. Now we will discuss reasons 2, 3, and 4.

6.3.1   Use transmorphic for arguments of overloaded functions

You want to implement a function such as `foo(x)`:

```
real scalar foo(real    vector x)
real scalar foo(string scalar x)
```

Notice that the function returns a `real` `scalar` regardless of how `x` is specified. `x` itself can be of different types, in this case, a `real` `vector` or a `string` `scalar`.

To code `foo()`, you begin by declaring argument x to be `transmorphic`:

```
real scalar foo(transmorphic x)
```

In the body of `foo()`, you write code to verify that x is as it is supposed to be, a **real vector** or a **string scalar**, and then you call separate subroutines to handle each case. We will write such a function in section 8.4.

6.3.2   Use transmorphic for output arguments

Sometimes functions need to return more than one thing. Mata's built-in function `eigensystem(A, X, L)` is an example. `eigensystem()` calculates the eigenvectors and eigenvalues of matrix A. Arguments X and L are provided to receive the calculated vectors and values. Whatever values X and L currently contain, if anything, are irrelevant. When `eigensystem()` returns, X will contain the eigenvectors and L will contain the eigenvalues.

The author of the function declared the output arguments to be `transmorphic`:

```
void eigensystem(numeric matrix A, transmorphic X,
                 transmorphic L)
{
        ...
}
```

It was important for the author to do that. The author was perhaps tempted to declare X and L using explicit types:

```
void eigensystem(numeric matrix A,
                 numeric matrix X, numeric rowvector L)
```

It would be logical to declare X and L in this way because X will be a **numeric matrix** and L will be a **numeric rowvector** when the function returns. The problem is that the declared types of arguments are a specification about what they contain at the time the function is called, not a statement about what they will contain when the function returns. It might well happen that X and L do not meet the restrictions at the time the function is called, and if so, that would result in an unnecessary error:

```
              eigensystem():   3202  x[0,0] found where rowvector
> required
          callerfunction():      -  function returned error
                  <istmt>:       -  function returned error
    r(3202);
```

The error is unnecessary because the function will redefine X and L.

Output-only arguments should be declared `transmorphic`.

6.3.2.1   Use transmorphic for passthru variables

The final valid use of **transmorphic** is for declaring passthru variables, or variables that pass through a routine but that the routine never examines or modifies. **db** in the following example is a passthru variable:

```
string scalar getdata(...)
{
        transmorphic    db
        .
        .
        db = open_database(...)
        .
        .
        X = extract_from_database(db, ...)
        .
        .
        close_database(db)
        .
        .
        return(...)
}
```

db is passthru in getdata() because getdata() never looks at db's contents or modifies them. In the above code, db is declared inside getdata(). db would also be passthru if it had been an argument to the function:

```
string scalar getdata(transmorphic db, ...)
{
        .
        .
        X = extract_from_database(db, ...)
        .
        .
        return(...)
}
```

Declaring db as **transmorphic** instead of what it really is has advantages. First and foremost, declaring it **transmorphic** ensures that you do not have to recompile getdata() if the db functions are updated and even if the update changes db itself. db might be a structure or a class, and the author might add, delete, or rearrange members. That does not matter because, when you compiled getdata(), Mata did not even know that db was a structure or a class.

Declaring db **transmorphic** also ensures that your programs do not accidentally modify db's contents. You cannot modify them because if you referred to db.mbr, Mata would issue an error when you attempted to compile the routines.

If you are the author of functions that provide passthru variables, you should document them as `transmorphic` and keep to yourself their actual types. This benefits you and your users. You can fix bugs and even add features, and users of your code will not need to recompile their functions unless they want to exploit the new features you added.

6.3.3   You must declare structures and classes if not passthru

I showed you examples of structures and classes in sections 3.4.2 and 3.4.3 during the tour. We will discuss them more in chapters 10 and 12. Right now, I want to clarify something I said in the previous section. I said that if you do not explicitly declare variables containing structures or classes as structures or classes, then you cannot access their members. It therefore follows that if you need to access the members, you must declare them as what they really are.

Which is to say, declarations may be optional in Mata, but they are not optional where structures and classes are concerned. If you do not explicitly declare them as structures or classes, they will be assumed to be `transmorphic`, meaning that you will not be able to access their members in your program.

6.3.4   How to declare pointers

We first discussed pointers in section 3.4.4 during the tour. Pointer is an eltype and can be combined with any orgtype, just as all other eltypes can be, so `p` can be declared as any of

```
pointer scalar      p
pointer vector      p
pointer rowvector   p
pointer colvector   p
pointer matrix      p
```

The declarations above are incomplete but allowed. They are incomplete because they do not specify what `p` points to. It will help if you think of `p` and `*p` as if they were separate variables. The above declarations are fully explicit as to `p`, but what is `*p`? You can specify the type of `*p` parenthetically:

```
pointer(real matrix) scalar   p
```

The above declaration states that `*p` is a real matrix. Pointers can point to other pointers, although that rarely occurs. If you have such a case, you can nest the parenthetical phrases:

```
pointer(pointer(real matrix) scalar) scalar   p
```

If you have a pointer to a pointer ... to a real matrix, you can continue nesting. I have never had occasion to use anything more than three-deep pointers:

```
pointer(pointer(pointer(real matrix) scalar) scalar) scalar p
```

When you omit the parenthetical phrase,

```
pointer scalar      p
```

Mata assumes that *p is a **transmorphic matrix**, just as Mata assumes for all variables with omitted declarations. I recommend specifying full declarations, although it is not necessary that you do so as long as *p is **real**, **complex**, **string**, or **pointer**.

It is necessary that you specify full declarations if *p is a structure or a class. More correctly, it is necessary unless *p is a passthru variable. Said differently, you will not be able to use the construct p->*mbr* unless you declare p to be

```
pointer(struct name scalar) scalar p
```

or

```
pointer(class name scalar) scalar p
```

7 Mata's strict option and Mata's pragmas

7.1 Overview

Mata has a strict setting to require that variables be explicitly declared. The setting also flags questionable constructs that might be mistakes. The strict setting is off by default.

Variable p2 is undeclared in the following program, but even so, Mata compiles the code:

```
: void quadraticroots(real scalar a, real scalar b,
> real scalar c)
> {
>         p2 = sqrt(b^2-4*a*c)
>         return( (-b+p2, -b-p2)/(2*a) )
> }
:
```

Had the strict setting been on, the program would not have compiled:

```
: void quadraticroots(real scalar a, real scalar b,
> real scalar c)
> {
>         p2 = sqrt(b^2-4*a*c)
>         return( (-b+p2, -b-p2)/(2*a) )
> }
variable p2 undeclared
r(3000);
```

Mata's strict setting flags questionable constructs, too. In the following code, variables X and L are used before they are initialized:

```
: real matrix mysubroutine(...)
> {
>          real matrix    result
>          real matrix    A, X, L
>
>              .
>              .
>              .
>          eigensystem(A, X, L)
>              .
>              .
>              .
>          return(result)
> }
note: variable X may be used before set
note: variable L may be used before set
```

Having uninitialized variables is usually a mistake, but not in this case. `eigensystem()` is a Mata library function that returns its results in X and L. You can tell the compiler that X and L are intentionally unset by using **pragma unset** *varname*:

```
: mata drop mysubroutine()

: real matrix mysubroutine(...)
> {
>          real matrix    result
>          real matrix    A, X, L
>
>              .
>              .
>              .
>          pragma unset X
>          pragma unset L
>          eigensystem(A, X, L)
>              .
>              .
>              .
>          return(result)
> }
```

A pragma is an instruction to the compiler. In this case, **pragma unset X** and **pragma unset L** instruct the compiler that X and L were intentionally left uninitialized.

When Mata produces warning messages such as "variable X may be used before set", you either fix the code or use pragmas to suppress the message. When the code does not need fixing, you suppress the message so that later, when the code is recompiled, you do not wonder all over again whether there is a mistake.

You place the pragmas directly above the questionable constructs, not at the top of the program. This way, if you later modify the code and use the uninitialized variable before the pragma appears, that will be flagged; this time, it might really be a mistake.

7.2   Turning matastrict on and off

Mata's strict setting is called `matastrict`, and it is off by default. You set `matastrict` using Stata's `set` command.

```
. set matastrict on
. ...
. set matastrict off
```

You can also set `matastrict` from inside Mata. Prefix the `set` command with `mata`:

```
. mata:
                                        ─── mata (type end to exit) ───
: mata set matastrict on
: ...
: mata set matastrict off
```

You do not need to turn `matastrict` off when compiling code in do- or ado-files because Stata automatically restores `matastrict` to its default setting.

7.3   The messages that matastrict produces, and suppressing them

`matastrict` produces the following messages:

```
variable _____ undeclared
note: variable _____ may be used before set
note: variable _____ unused
note: argument _____ unused
note: variable _____ set but not used
```

The first message is an error, and the remaining messages are warnings.

Even though the example I showed in the introduction was a false positive, warnings usually flag real problems, such as in the following routine:

```
: void quadraticroots(real scalar a, real scalar b,
                       real scalar c)
> {
>         real scalar p2
>
>         return( (-b+p2, -b-p2)/(2*a) )
> }
note: variable p2 may be used before set
note: argument c unused
```

In this case, the warning messages appear because I forgot to code the line

```
p2 = sqrt(b^2-4*a*c)
```

I have forgotten to code lines, or have accidentally deleted them, more often than I care to admit.

Mata reports "variable _____ unused" when variables are declared but unused. Here is an example:

```
: real matrix weights(...)
> {
>          real scalar  N, k
>          real matrix  W, V
>
>              .
>              .
>              .
>          return(W)
> }
note: variable k unused
```

Either the declaration of variable **k** needs to be deleted or there is a substantive error in the body of the program. Perhaps, just as I did, you forgot to code a line.

The possibilities are the same with the "variable _____ set but not used" message. In the following program, **k** is declared, it is set to 0, but after that it is never referred to again:

```
: real matrix weights(...)
> {
>          real scalar  N, k
>          real matrix  W, V
>
>              .
>              .
>              .
>          k = 0
>              .
>              .
>              .
>          return(W)
> }
note: variable k set but not used
```

Either use **k** or delete it.

The message "argument _____ unused" is a variation of "variable _____ unused". Mata distinguishes between arguments and variables because sometimes there are good reasons to have unused arguments. In the following code, argument m is unused:

```
: real matrix weights(..., real scalar m, ...)
> {
>         real scalar  N
>         real matrix  W, V
>
>              .
>              .
>              .
>         return(W)
> }
note: argument m unused
```

Although the obvious solution is to delete m, there might be good reasons for keeping it. The variable might have to do with a feature you are planning to add. In that case, you might retain the argument so that callers are forced to pass it in preparation for that happy event. Nevertheless, until the feature is added, you should add code to verify that m has the appropriate value given what **weights()** currently returns:

```
: real matrix weights(..., real scalar m, ...)
> {
>         real scalar  N
>         real matrix  W, V
>
>         if (m!=1) {
>                 _error("m currently must be 1")
>         }
>              .
>              .
>              .
>         return(W)
> }
```

The warning message no longer appears because m is now used by the line of code verifying that it equals 1. Today, **weights()** returns type m==1 matrices. Perhaps someday it will also return type m==2 matrices, matrices calculated using a different formula.

The above is a better solution than using a pragma to suppress the warning message, such as

```
: real matrix weights(..., real scalar m, ...)
> {
>          real scalar  N
>          real matrix  W, V
>
>          pragma unused m    // argument for future development
>          .
>          .
>          .
>          return(W)
> }
```

Here the warning message no longer appears because we coded **pragma unused m**.

In this case, it is dangerous to use a pragma because today's users might specify m incorrectly given what you intend for the future, yet they would have no way of detecting their mistake. Nonetheless, I have been known to use this approach: Imagine that **weights()** has another argument, Z, a matrix that might be symmetric. The formulas implemented today are the nonsymmetric formulas, which work when Z is symmetric, too. Someday, I plan to add code to **weights()** that will calculate W more quickly when Z is symmetric. Thus, I suggest that users specify m indicating whether Z is symmetric, and, someday, calculations with symmetric Z matrices might run more quickly.

8 Mata's function arguments

8.1 Introduction

This chapter concerns function arguments.

First, Mata functions can change the values stored in the arguments they receive from callers. Changing them changes the values stored in the caller's variables, so you must exercise caution not to change the values unintentionally. Intentionally changing them, however, can be useful. It is a way of returning multiple results.

Second, Mata functions can have a varying number of arguments. There is no downside here and it is easy to program.

Third, Mata functions can have multiple syntaxes. This involves combining a varying number of arguments with the `transmorphic` variable type.

8.2 Functions can change the contents of the caller's arguments

Consider a function such as

```
real scalar foo(real matrix A, real matrix B)
```

and a caller of `foo()` who codes

```
result = foo(mat1, mat2)
```

121

Mata is a call-by-reference language, meaning that `mat1` and `mat2` are literally the `A` and `B` that `foo()` receives. If `foo()` changes `A`, then `mat1` is changed because `A` and `mat1` are the same variable. The same applies to `B` and `mat2`.

Here is a demonstration of this with a one-argument function:

```
: void goo(xvar)
> {
>            xvar = I(2)
> }
: X
  1
: goo(X)
: X
[symmetric]
        1   2
    1 | 1
    2 | 0   1
```

From the caller's perspective, `X` had 1 in it, and after calling `goo()`, `X` was changed to be a 2 x 2 identity matrix.

Most official functions in Mata do not change what is stored in the caller's variables. That is by intent, of course, and the programmers who wrote the functions had to be careful not to change the arguments by accident.

Other Mata functions do change the caller's variables. Mata's `eigensystem()` function calculates the eigenvectors and eigenvalues of a matrix `A`. The function needs to return both the vectors (a matrix) and the values (a vector). `eigensystem()` returns the two results in arguments that the caller provides. To obtain results for matrix `A`, the caller calls `eigensystem(A, X, L)`. The eigenvectors are returned in `X`, and the eigenvalues are returned in `L`. Meanwhile, the function itself returns void.

There is no deep programming issue that needs explaining, although it is perhaps worth repeating that arguments `X` and `L` should be declared as **transmorphic**. The code is

```
void eigensystem(numeric matrix A, transmorphic X,
                 transmorphic L)
{
        .
        .
        X = ...
        L = ...
        .
        .
        return()
}
```

X and L should be declared `transmorphic` because `eigensystem()` does not care about the type of result currently stored in them. In the assignment to them, `eigensystem()` will store the appropriate real or complex result, as explained in section 6.3.2.

8.2.1 How to document arguments that are changed

Communicating to callers when functions change arguments is obviously important. Less obviously, it is important that you communicate it to yourself, and do so before writing the code. I include a comment above routines that I write showing which arguments are inputs and which are outputs. I put bars over outputs and lines under inputs. Mata's `sqrt()` function looks like this:

```
/*
            _____
      result = sqrt(x)
                    _
*/
```

The diagram says the `sqrt()`'s argument remains unchanged.

Most functions look like `sqrt()`. The diagram for Mata's `eigensystem()` function, however, is

```
/*
                      _   _
      (void) eigensystem(A, X, L)
                        _
*/
```

This diagram says that A is unchanged and X and L are changed, and it says something else. It says that the function does not care what was stored in X and L. That information is ignored.

Mata's `invsym()` function has a diagram identical to that of `sqrt()`:

```
/*
            _____
      Ainv = invsym(A)
                    _
*/
```

There is another version of `invsym()`, however, that has a rather surprising diagram:

```
/*
                        _
      (void)   _invsym(A)
                        _
*/
```

Can you interpret it? A is underlined, which means that the contents of A are used by the function; A has a line above it, which means that the contents of A are changed by the function; and so there is only one possibility: `_invsym(A)` replaces the input matrix with its inverse. Having arguments that are both inputs and outputs is usually best avoided, but in this case it is justified by the memory that is conserved.

So let's consider function syntaxes. Mata's library routines follow reasonably good style.

Most work like this:

No arguments used to return values:

```
_____
result = fcn(A,  B,  C,  ...)
              -   -   -
```

Others work like this:

Return values in arguments in rightmost or leftmost positions:

```
                  _   _   ___
(void)   fcn(A,  B,  C,  D,  E,  ...)
              -   -   -
             _   _
(void)   fcn(A,  B,  C,  D,  E,  ...)
              -   -   -   ---
```

Still others work like this:

Return values in arguments plus a returned value indicating success or failure:

```
_____           _   _   ___
success = fcn(A,  B,  C,  D,  E,  ...)
                   -   -   -
_____      _   _
success = fcn(A,  B,  C,  D,  E,  ...)
                   -   -   -   ---
```

A few work like this:

Input changed to output:

```
             _   _   ___
(void) _fcn(A,  B,  ...)
             -   -   ---
```

None work like this:

Return values randomly interspersed with input values

```
             _         _        ???
(void)   fcn(A,  B,  C,  D,  ...)
             -   -        -      ???
```

What makes good style is predictability and adjacency. It should be predictable when the function will change the contents' arguments, and the changed arguments should be adjacent.

In code you write, even if it is just for yourself, do not change the contents of arguments without good reason. When there is good reason, make it clear which arguments are modified. Make it clear both in your code and in any documentation that you write.

8.2.2   How to write functions that do not unnecessarily change arguments

If you do not intend to change the caller's variables, make sure that you do not change them. For instance, assume that you are writing

```
real matrix myfcn(real matrix A, real scalar i)
{
        . . .
}
```

As you write the code, you realize that it would be convenient if you added 1 to i. Doing that, however, would change the caller's variable. The right way to code such cases is to make a copy of the caller's variable and increment it.

```
real matrix myfcn(real matrix A, real scalar caller_i)
{
        real scalar     caller_i
        .
        .
        i = caller_i + 1
        .
        .
}
```

8.3   How to write functions that allow a varying number of arguments

Mata functions have a fixed number of arguments by default, but you can write functions that allow the number to vary. Consider a function `matrixmanip()`, in which we want to make the second argument optional:

```
R = matrixmanip(A [, tolerance])
```

We plan to make the default `tolerance` 1.0x–1a, which is roughly 1.0e–8 as explained in section 5.2.1.2. Here is how we make `matrixmanip()` do that:

```
real matrix matrixmanip(real matrix A |, real scalar tolerance)
{
        .
        .
        if (args()==1) tolerance = 1.0x-1a
        .
        .
}
```

The vertical bar in front of the comma separates required arguments from optional ones. Arguments to the right of the bar are optional. The line

```
if (args()==1) tolerance = 1.0x-1a
```

is how we set **tolerance** if the second argument is not specified. Built-in function **args()** returns the number of arguments the caller specified.

Regardless of whether the caller specifies one or two arguments, Mata makes the situation look as if two arguments were specified. Mata itself provides the second argument if necessary and sets the argument to missing value in the case of **tolerance** because it was declared to be a scalar. If **tolerance** had been declared to be a vector or a matrix, Mata would have made it 0×1, 1×0, or 0×0.

Thus, either the user specified the second argument or Mata did. Mata function **args()** returns the number of arguments that the user specified. Because **matrixmanip()** allows one or two arguments, we know **args()** must be 1 or 2. If **args()==2**, then the caller specified the tolerance. If **args()==1**, then **matrixmanip()** executes code to set **tolerance** to the default:

```
if (args()==1) tolerance = 1.0x-1a
```

I previously warned against changing values stored in arguments unnecessarily, but filling in **tolerance** when Mata specifies it is okay. Mata does not care if we change it, and in fact, Mata expects us to change it.

The code for handling multiple required and optional arguments is similar to the above code. In the following code, the first three arguments are required and the last two are optional. That means the user could specify three, four, or five arguments.

```
real matrix matrixmanip(real matrix A, real vector w,
                        real matrix B
                    |, real scalar tol1, real scalar tol2)
{
        .
        .
        if (args()<4) tol1 = 1.0x-1a
        if (args()<5) tol2 = 1.0x-1a
        .
        .
}
```

Note the cuteness of the code that sets the defaults. `tol1` is set if one, two, or three arguments were specified. `tol2` is set if one, two, three, or four arguments were specified.

8.4 How to write functions that have multiple syntaxes

Say you wish to write a function that does different things with different types of arguments. For example, you wish to write with two syntaxes:

```
real scalar foo(real vector v)
real scalar foo(string scalar s)
```

This example might seem absurd, but it is not. Perhaps `foo()` calculates a kind of check sum. Whatever it is that `foo()` does, the solution is to use **transmorphic** for the argument, diagnose whether the argument is **real** or **string**, and call an appropriate subroutine:

```
real scalar foo(transmorphic x)
{
        if (eltype(x)=="real")   return(foo1(x))
        if (eltype(x)=="string") return(foo2(x))

        _error("real or string required")
}
real scalar foo1(real vector v)
{
        .
        .
}
real scalar foo2(string scalar s)
{
        .
        .
}
```

Function `foo()` calls `foo1()` or `foo2()`, and those functions include the appropriate declaration.

Functions with multiple syntaxes sometimes allow different numbers of arguments, too. Perhaps `foo()` needs two arguments when a real vector is specified. Imagine that the syntax we need to implement is

```
real scalar foo(real vector v, real scalar mesh)
real scalar foo(string scalar s)
```

The code could read

```
real scalar foo(transmorphic one |, transmorphic two)
{
        if (eltype(x)=="real") {
                if (args()!=2) _error(3001)
                return(foo1(one, two))
        }
        if (eltype(x)=="string") {
                if (args()!=1) _error(3001)
                return(foo2(one))
        }
        _error("real or string required")
}
real scalar foo1(real vector v, real scalar mesh)
{
        .
        .
        .
}
real scalar foo2(string scalar s)
{
        .
        .
        .
}
```

I used **_error(3001)** to produce the standard Mata error message for the incorrect number of arguments specified; see **help m2 errors**.

9 Programming example: n_choose_k() three ways

9.1 Overview

We are now ready to use Mata as a serious programmer would use it, or more correctly, as three different serious programmers would use it. We are going to write a Mata routine and package it three ways.

We will package it as a do-file that can be used from other do-files.

We will package it as an ado-file, creating a new Stata command.

We will package it as a Mata library routine so that it can be used in other Mata and Stata code just as if it were originally a part of Mata.

We first need to develop a routine worthy of this effort.

9.2   Developing n_choose_k()

We are going to write a routine to calculate the number of combinations of n things chosen k at a time. The formula for the number of combinations is

$$c = \frac{n!}{(n-k)!\, k!}$$

I am sure you can imagine writing code to calculate c. Here is an obvious routine using Mata's built-in factorial function:

```
real scalar choose(real scalar n, real scalar k)
{
        return(factorial(n) / ( factorial(n-k)*factorial(k) ))
}
```

The code is clear, concise, and produces accurate results, and yet the code would be unacceptable if turned in by a StataCorp programmer. The problem is that choose(1000, 1) returns missing value when it should have returned 1,000. How many ways can you choose 1 thing from 1,000? You could choose the first thing, the second, ..., or the thousandth. There are 1,000 ways.

The code produced missing value because it attempted to calculate 1,000 factorial. That result is too large to be calculated on a digital computer, being approximately $4.024 \times 10^{2,567}$. The largest number that the computer can store is approximately 8.988×10^{307}. A better routine would perform the calculation in the logs by using Mata's lnfactorial() function:

```
real scalar choose2(real scalar n, real scalar k)
{
        return(exp(
                lnfactorial(n) -
                    (lnfactorial(n-k)+lnfactorial(k))
                )
        )
}
```

According to this routine, there are 1000.000000000994532 ways to choose 1 thing from 1,000. Relative to the true answer, the error is only 9.945e–13, so I view the result as acceptable. We could nevertheless improve the accuracy if we rounded the calculated result to the nearest integer by using Mata's round() function:

```
real scalar choose3(real scalar n, real scalar k)
{
        return(round(
                        exp(lnfactorial(n) -
                                (lnfactorial(n-k)+lnfactorial(k))
                        )
                )
        )
}
```

With this change, `choose3()` reports that there are exactly 1,000 ways to choose 1 thing from 1,000.

This function is still not perfect. It calculates the correct result for 1 thing chosen from 10,000,000, but not for 1 thing chosen from 100,000,000. It produces 99,999,978, which is off by 22, a relative error of 2.20e–7.

Can we write a routine that does better? Let's reconsider the mathematical formula

$$c = \frac{n!}{(n-k)!\ k!}$$

Computers can calculate formulas like the one above with near-perfect accuracy because they can multiply and divide without loss of precision, by which I mean that if a computer calculates `a=b/c`, then it will be true that `a*c==b`. Lots of people think computers are accurate, and they are right when the operators are multiplication and division. Other people, knowing computers perform finite-precision arithmetic, think computer inaccuracy is equally shared across all operators. It is not. Round-off error arises because of addition and subtraction, not multiplication and division. Thus, I can tell at a glance that

$$c = \frac{n!}{(n-k)!\ k!}$$

can be calculated on a computer more accurately than

$$c = \exp(\mathrm{lnfactorial}(n-k) - \mathrm{lnfactorial}(n-k) - \mathrm{lnfactorial}(k))$$

I can tell at a glance because the first equation involves multiplication and division, while the second involves subtraction. Both equations share whatever error might be introduced by the calculation of $n-k$, and because n and k are integers, that is no error at all as long as n, k, and $n-k$ are in the range $\pm 9,007,199,254,740,992$.

My complaint when we implemented code for the factorial formula was that the code produced missing values due to overflow in cases where the result was a reasonable number. I focused on the example $n = 1000$ and $k = 1$, but there are many other examples I could have mentioned. It turns out there is way we could have avoided the $n!$ calculation that caused the overflow. If I asked you to calculate c by hand for $n = 1000$ and $k = 1$, you would proceed like this:

$$c = \frac{1000!}{(1000 - 1)! \, 1!}$$

$$= \frac{1000!}{999!}$$

You would first write out the factorial calculation, and then cancel the 999 in the numerator with the 999 in the denominator, cancel the 998s, cancel the 997s, and so on. You would have written

$$\frac{1000!}{999!} = \frac{1000 \times 999 \times 998 \times \cdots \times 1}{999 \times 998 \times \cdots \times 1}$$

$$= 1000$$

Working problems by hand and then writing code that mimics what you did is an excellent way to develop computer routines. Here you just calculated $n!(n - k)!/k!$ by using

$$c = \frac{n \times (n - 1) \times \cdots \times (n - k)}{k!}$$

The above formula saves time for $n = 1000$ and $k = 1$, but if I had asked you to calculate c for $n = 1000$ and $k = 999$, you would have canceled numbers in the numerator with numbers due to $(n - k)!$ in the denominator instead of $k!$.

$$c = \frac{n \times (n - 1) \times \cdots \times k}{(n - k)!}$$

The solution written in Mata code is

```
real scalar n_choose_k(real scalar n, real scalar k)
{
        return( n-k > k ?
                    nfactorial_over_kfactorial(n, n-k) /
                    factorial(k)
                :
                    nfactorial_over_kfactorial(n,   k) /
                    factorial(n-k)
                )
}
```

We will need to write the **nfactorial_over_kfactorial()** subroutine, but that is a detail. Our insight—cancellation of like terms in the numerator and the denominator—is codified in **n_choose_k()**.

For no good reason, I am going to rewrite **n_choose_k()** because I would like to substitute calls to **nfactorial_over_kfactorial()** for calls to **factorial()**:

```
real scalar n_choose_k(real scalar n,  real scalar k)
{
        return( n-k > k ?
                        nfactorial_over_kfactorial(n, n-k) /
                        nfactorial_over_kfactorial(k,    1)
              :
                        nfactorial_over_kfactorial(n,    k) /
                        nfactorial_over_kfactorial(n-k, 1)
              )
}
```

Calling **nfactorial_over_kfactorial**(x, 1) is another way to calculate $x!$, and it appealed to me to use our subroutine instead of **factorial()**. As I said, I did this for no good reason. Here is our subroutine:

```
real scalar nfactorial_over_kfactorial(real scalar n,
                                       real scalar k)
{
        real scalar     result, i
        if (n<0 | n>1.0x+35 | n!=trunc(n)) return(.)
        if (k<0 | k>1.0x+35 | k!=trunc(k)) return(.)

        result = 1
        for (i=n; i>k; --i) result = result*i

        return(result)
}
```

I have just one comment about this code, and it concerns the lines

```
if (n<0 | n>1.0x+35 | n!=trunc(n)) return(.)
if (k<0 | k>1.0x+35 | k!=trunc(k)) return(.)
```

The two lines return missing value when **n** or **k** is less than 0, or is greater than **1.0x+35**, or is not an integer. Function **trunc**(x) returns the largest integer less than or equal to x. As for the **1.0x+35**, it is a shorter way of writing 9,007,199,254,740,992. I mentioned above that integers in the range $\pm$**1.0x+35** are not subject to rounding, although when I originally said it, I used the ungainly base-10 number.

You might suspect that I should have added a third line, namely,

```
if (n>=.| k>=.) return(.)
```

Adding the line is unnecessary because missing value is greater than all nonmissing values, and the first two lines already return missing value when n or k is greater than **1.0x+35**.

n_choose_k() and its subroutine **nfactorial_over_kfactorial()** are the routines we are going to package as a do-file, as an ado-file, and as a Mata library routine.

9.3   n_choose_k() packaged as a do-file

Do-files tend to be written in ways peculiar to the problem at hand, and that means there is no best single template to show you. I will show you the do-file I wrote back when I was writing chapter 4, where **n_choose_k()** made a brief appearance. I did not want to risk showing you code that did not work, so I wrote a do-file containing the code and verifying that it produced the correct answers.

As I said, do-files are peculiar to the problem at hand, and I had my own particular problem. I will show you my do-file, and then I will show you a more general approach.

9.3.1   How I packaged the code: n_choose_k.do

The do-file I originally wrote had one goal: to prove that **n_choose_k()** produced correct results. I put the code and tests of the code into one file, which I named **n_choose_k.do**.

```
                                                          n_choose_k.do
 1:  version 15
 2:  set matastrict on
 3:
 4:  mata:
 5:
 6:  /*          _
 7:          N = n_choose_k(n, k)
 8:                      -   -
 9:
10:                  N = n! / ( (n-k)! k! )
11:
12:          Formula evaluated using canceling to increase domain.
13:  */
14:
15:  real scalar n_choose_k(real scalar n, real scalar k)
16:  {
17:          return( n-k > k ?
18:                  nfactorial_over_kfactorial(n, n-k) /
19:                  nfactorial_over_kfactorial(k, 1)
20:                      :
21:                  nfactorial_over_kfactorial(n, k) /
22:                  nfactorial_over_kfactorial(n-k, 1)
23:                  )
24:  }
25:
26:  /*          _____
27:          result = nfactorial_over_kfactorial(n, k)
28:                                              -   -
29:
30:          Returns n!/k! calculated as n*(n-1)*...*(k+1).
31:          Note: 0! = 1.
32:  */
33:
34:
35:  real scalar nfactorial_over_kfactorial(real scalar n, real scalar k)
36:  {
37:          real scalar     result, i
38:
```

```
39:            if (n<0 | n>1.0x+35 | n!=trunc(n))   return(.)
40:            if (k<0 | k>1.0x+35 | k!=trunc(k))   return(.)
41:
42:            result = 1
43:            for (i=n; i>k; --i) result = result*i
44:
45:            return(result)
46:  }
47:  end
48:
49:
50:  mata:
51:
52:  // Test 1
53:  //   For small n, n_choose_k(n, k) should produce the same result
54:  //   as directly calculating n!/((n-k)!k!).
55:
56:  real scalar nk(real scalar n, real scalar k)
57:   return(factorial(n) / (factorial(n-k)*factorial(k)))
58:
59:  for (n=1; n<=5; n++) {
60:   for (k=0; k<=n+1; k++) {
61:   /* printf("Ch(%f, %f) = %f\n", n, k, n_choose_k(n,k)) */
62:   assert(n_choose_k(n, k) == nk(n, k))
63:   }
64:  }
65:
66:  assert(n_choose_k(5,0) == 1)
67:  assert(n_choose_k(5,6) == .)
68:
69:  // Test 2
70:  //   n_choose_k() should produce results when direct calculation
71:  //   fails.
72:
73:  assert(n_choose_k(1000,    1) == 1000)
74:  assert(n_choose_k(1000,    2) == 1000*999/2)
75:  assert(n_choose_k(1000, 1000) ==    1)
76:  assert(n_choose_k(1000,  999) == 1000)
77:  assert(n_choose_k(1000,  998) == n_choose_k(1000, 2))
78:  assert(        nk(1000,    1) ==    .)   // sic, FYI
79:  assert(        nk(1000, 1000) ==    .)   // sic, FYI
80:
81:  // Test 3
82:  //   Missing values.
83:
84:  assert(n_choose_k(-1,   1)==.)
85:  assert(n_choose_k( 1,  -1)==.)
86:  assert(n_choose_k( 5,   .)==.)
87:  assert(n_choose_k( .,   2)==.)
88:  assert(n_choose_k( .,   .)==.)
89:
90:  end
```
——————————————————————————————————————— n_choose_k.do ———————

Note the comments in the file. When I write code, I include comments. It has been only a few weeks since I wrote the do-file, and I am already glad that I have comments. They reminded me how the code works and made it easier for me to write this chapter. If I ever need to modify the code, the comments will help then, too.

With or without comments, the do-file solved my problem. To verify that **n_choose_k()**
worked, all I had to do was type **do n_choose_k**:

```
. do n_choose_k
(output omitted)

. _
```

Because the do-file ran without error, I know that **n_choose_k()** produces correct answers for the particular problems that I ran. I do not need to look at the output, because
I wrote the tests using Mata's built-in **assert()** function. One of the lines in the file is

```
assert(n_choose_k(1000,    1) == 1000)
```

assert(*expr*) does nothing if *expr* is true. If it is not true, however, **assert()** aborts
with error. If **n_choose_k()** had a bug and **n_choose_k(1000, 1)** did not equal 1,000,
here is what would have happened:

```
. do n_choose_k
(some output omitted)
: assert(n_choose_k(1000,    1) == 1000)
                        assert():  3498   assertion is false
                        <istmt>:      -   function returned error
(14 lines skipped)
```
```
r(3498);

end of do-file

r(3498);


. _
```

The do-file would have stopped at the first unexpected result, and that is difficult to
ignore. You can read more about **assert()** by typing **help mata assert()**. You can
even see **assert()**'s source code by typing **viewsource assert.mata**.

Some programmers would have written the certification part of my do-file to display
the results instead of using **assert()**. That is, instead of lines such as

```
assert(n_choose_k(1000,    1) == 1000)
assert(n_choose_k(1000,    2) == 1000*999/2)
assert(n_choose_k(1000, 1000) ==    1)
assert(n_choose_k(1000,  999) == 1000)
```

they would have coded the lines

```
n_choose_k(1000,    1)
n_choose_k(1000,    2)
n_choose_k(1000, 1000)
n_choose_k(1000,  999)
```

The output from their do-file would certainly have been more entertaining:

```
: n_choose_k(1000, 1)
  1000

: n_choose_k(1000, 2)
  499500

: n_choose_k(1000, 1000)
  1

: n_choose_k(1000, 999)
  1000
```

Are the answers right? Determining that would require looking at and thinking about the results presented. Would you notice if n_choose_k(1000,2) produced 498,500 instead of 499,500? I ran 250 tests in my do-file. If one or two produced incorrect results, would you have spotted them? How long would you need to review the output? It took me about 15 minutes to code the asserted results.

My tests produced no output, but I have no objections to tests that do. Notice that in one part of the do-file, there is a commented-out **printf()** line:

```
for (n=1; n<=20; n++) {
        for (k=0; k<=n+1; k++) {
                // printf("Ch(%f, %f) = %f\n", n, k,
                          n_choose_k(n,k))
                assert(n_choose_k(n, k) == nk(n, k))
        }
}
```

The first draft of **n_choose_k()** had a bug, and I needed output to find it. I later commented out the **printf()** statement, but there is no requirement or rule that I do that.

9.3.2 How I could have packaged the code

My do-file served my purposes in writing this book. If I had wanted to package n_choose_k() for use by other do-files, however, I would have written the do-file differently. I would have created two do-files:

n_choose_k.mata	do-file containing function definitions
test_n_choose_k.do	do-file containing tests that functions produce correct results

File **n_choose_k.mata** would have defined the functions.

```
──────────────────────────────────────── n_choose_k.mata ─────────
version 15
set matastrict on

mata:
definitions of functions appear here
end
──────────────────────────────────────── n_choose_k.mata ─────────
```

File **test_n_choose_k.do** would have loaded the functions and tested them.

```
──────────────────────────────────────── test_n_choose_k.do ─────────
// version number intentionally omitted

clear all
run n_choose_k.mata

mata:
code to test functions appears here
end
──────────────────────────────────────── test_n_choose_k.do ─────────
```

Developing the code and tests would have been just as easy with two files as with one. File **test_n_choose_k.do** would have started with only a few lines and no tests at all:

```
──────────────────────────────────────── test_n_choose_k.do ─────────
// version number intentionally omitted

clear all
run n_choose_k.mata
──────────────────────────────────────── test_n_choose_k.do ─────────
```

As I wrote the code in **n_choose_k.mata**, I would have typed

```
. do test_n_choose_k
```

to see whether the code in the file I was writing had compile-time errors or warning messages. Once the code compiled without errors, I would have set about adding tests to **test_n_choose_k.do**, such as

```
──────────────────────────────────────── test_n_choose_k.do ─────────
// version number intentionally omitted

clear all
run n_choose_k.mata

mata:
assert(n_choose_k(1000,    1) == 1000)
end
──────────────────────────────────────── test_n_choose_k.do ─────────
```

I would run the file,

```
. do test_n_choose_k
  (output omitted)
```

and then add another test:

```
────────────────────────────────── test_n_choose_k.do ──────────
// version number intentionally omitted

clear all
run n_choose_k.mata

mata:
assert(n_choose_k(1000,    1) == 1000)
assert(n_choose_k(1000,    2) == 1000*999/2)
end
────────────────────────────────── test_n_choose_k.do ──────────
```

And so I would continue.

When I am done, and the code is written and works to my satisfaction, I do not throw away test_n_choose_k.do. It is my completed certification file, which I can use to test n_choose_k() and its subroutines in the future. Why would I want to do that? Because a year from now, a reader might report a bug in n_choose_k(), and I have been known to fix one bug just to introduce another. Typing do test_n_choose_k will provide evidence that my fix has not made things worse. Setting myself up for the next bug, I will add code to test_n_choose_k.do that shows I fixed this bug.

In the meantime, I can use n_choose_k.mata in my analysis do-files:

```
────────────────────────────────────── myanalysis1.do ──────────
version 15
clear all
.
.
.
run n_choose_k.mata
.
.
.
use n_choose_k() function
.
.
────────────────────────────────────── myanalysis1.do ──────────
```

9.3.2.1   n_choose_k.mata

We at StataCorp use file extension .mata for files that contain Mata functions, structures, and class definitions. If a do-file ends in .mata, we know that it defines new Mata functions and the like, but it does not clear all or otherwise disturb Stata's or Mata's environment.

The template for `.mata` files is

```
──────────────────────────────────────── name.mata ───────────
version #
set matastrict on

mata:
definitions of functions, structures, and classes appear here
end
──────────────────────────────────────── name.mata ───────────
```

The two lines at the top of the file are important.

The first line states the Stata version at the time the code was written. It ensures that the code will be compiled with newer versions of Stata just as it was originally compiled.

The second line tells Mata to impose strict standards when compiling the code. Why we want strict standards was explained in chapter 7.

Here is the **n_choose_k.mata** do-file:

```
──────────────────────────────────────── n_choose_k.mata ───────────
 1:  version 15
 2:  set matastrict on
 3:
 4:  mata:
 5:
 6:  /*         _
 7:          N = n_choose_k(n, k)
 8:                          -  -
 9:
10:                    N = n! / ( (n-k)! k! )
11:
12:          Formula evaluated using canceling to increase domain.
13:  */
14:
15:  real scalar n_choose_k(real scalar n, real scalar k)
16:  {
17:          return( n-k > k ?
18:                  nfactorial_over_kfactorial(n, n-k) /
19:                  nfactorial_over_kfactorial(k, 1)
20:                  :
21:                  nfactorial_over_kfactorial(n, k) /
22:                  nfactorial_over_kfactorial(n-k, 1)
23:                  )
24:  }
25:
26:  /*         _____
27:          result = nfactorial_over_kfactorial(n, k)
28:                                                -  -
29:
30:          Returns n!/k! calculated as n*(n-1)*...*(k+1).
31:          Note: 0! = 1.
32:  */
33:
34:
```

```
35:   real scalar nfactorial_over_kfactorial(real scalar n, real scalar k)
36:   {
37:           real scalar     result, i
38:
39:           if (n<0 | n>1.0x+35 | n!=trunc(n))  return(.)
40:           if (k<0 | k>1.0x+35 | k!=trunc(k))  return(.)
41:
42:           result = 1
43:           for (i=n; i>k; --i) result = result*i
44:
45:           return(result)
46:   }
47:   end
```
──────────────────────────────────── n_choose_k.mata ─────────

9.3.2.2   test_n_choose_k.do

Files named **test_***name*.**do** test that the code in *name*.**mata** works.

The template for **test_***name*.**do** is

──────────────────────────────────── test_*name*.do ─────────
```
// version number intentionally omitted
clear all
run name.mata

mata:
code to test functions appears here
end
```
──────────────────────────────────── test_*name*.do ─────────

So many do-files start with a **version** statement that we at StataCorp include a comment when we intentionally omit it. In this case, we omit it because once the functions
are compiled, they should work with the current and future versions of Stata.

Here is file `test_n_choose_k.do`:

```
                                                    test_n_choose_k.do
 1:  // version number intentionally omitted
 2:
 3:  clear all
 4:  run n_choose_k.mata
 5:
 6:  mata:
 7:
 8:  // Test 1
 9:  //   For small n, n_choose_k(n, k) should produce the same result
10:  //   as directly calculating n!/((n-k)!k!).
11:
12:  real scalar nk(real scalar n, real scalar k)
13:   return(factorial(n) / (factorial(n-k)*factorial(k)))
14:
15:  for (n=1; n<=5; n++) {
16:   for (k=0; k<=n+1; k++) {
17:   /* printf("Ch(%f, %f) = %f\n", n, k, n_choose_k(n,k)) */
18:   assert(n_choose_k(n, k) == nk(n, k))
19:   }
20:  }
21:
22:  assert(n_choose_k(5,0) == 1)
23:  assert(n_choose_k(5,6) == .)
24:
25:  // Test 2
26:  //   n_choose_k() should produce results when direct calculation
27:  //   fails.
28:
29:  assert(n_choose_k(1000,    1) == 1000)
30:  assert(n_choose_k(1000,    2) == 1000*999/2)
31:  assert(n_choose_k(1000, 1000) ==     1)
32:  assert(n_choose_k(1000,  999) == 1000)
33:  assert(n_choose_k(1000,  998) == n_choose_k(1000, 2))
34:  assert(         nk(1000,    1) ==     .)   // sic, FYI
35:  assert(         nk(1000, 1000) ==     .)   // sic, FYI
36:
37:  // Test 3
38:  //   Missing values.
39:
40:  assert(n_choose_k(-1,   1)==.)
41:  assert(n_choose_k( 1,  -1)==.)
42:  assert(n_choose_k( 5,   .)==.)
43:  assert(n_choose_k( .,   2)==.)
44:  assert(n_choose_k( .,   .)==.)
45:  end
                                                    test_n_choose_k.do
```

Seeing these files in their final form is misleading. In earlier drafts, `n_choose_k.mata` was not error free, and neither was `test_n_choose_k.do`. Let me tell you a little about the process that produced the two files.

You know that I did not use this two-file setup when I originally wrote the code, but I am going to tell you the story as if I did. The story varies from the truth in only a few, unimportant details.

I debugged **n_choose_k.mata** as I wrote `test_n_choose_k.do`. File `test_n_choose_k.do` started out like this:

```
——————————————————————————— test_n_choose_k.do ————————
// version number intentionally omitted
clear all
run n_choose_k.mata
——————————————————————————— test_n_choose_k.do ————————
```

I ran **test_n_choose_k.do** to discover that **n_choose_k.mata** had errors so severe that the code would not even compile. I fixed the mistakes, and once the code did compile, I added the first test to **test_n_choose_k.do**:

```
——————————————————————————— test_n_choose_k.do ————————
// version number intentionally omitted
clear all
run n_choose_k.mata
mata:
assert(n_choose_k(1000,     1) == 1000)
end
——————————————————————————— test_n_choose_k.do ————————
```

`assert()` reported that the assertion was false, of course. I found and fixed the bug (two of them) in **n_choose_k.mata** and ran **test_n_choose_k.do** again. It worked. Then I added another test:

```
——————————————————————————— test_n_choose_k.do ————————
// version number intentionally omitted
clear all
run n_choose_k.mata
mata:
assert(n_choose_k(1000,     1) == 1000)
assert(n_choose_k(1000,     2) == 1000*999/2)
end
——————————————————————————— test_n_choose_k.do ————————
```

That worked too. Feeling braver, I added more tests between runs. All went well for a while, but then `assert()` squawked. This happened in one of the more complicated tests where the test itself was executing a loop, and the error was not in the code but in the test itself! Both code and tests have to be debugged.

Debugging is something we all must do. Would you believe me if I told you that debugging sometimes takes less time if you do it while writing a test file? That really does happen. It happens when you fix one bug but the fix itself introduces a new one. After you fix a bug, you rerun the test file, and sometimes an earlier test fails because of a new bug you just introduced. If you were testing the code interactively, you would not rerun the tests you previously typed. It would be too tedious, and anyway, you would not even remember them.

It is just a little more work to type tests into a file than to type them interactively. You have to debug the code, so why not write a test file simultaneously?

9.3.3   Certification files

Let's review. I started with one do-file that contained code and tests, and I converted it into two do-files. Did I introduce errors doing that? It would have been easy to do. Whenever you pass a file through a text editor, you can accidentally make changes to it. Characters get added or deleted; entire lines can be deleted. You name it, and it has happened to me too. File `test_n_choose_k.do` has a wonderful feature. To verify that all is well, all I need to do is run it.

```
. do test_n_choose_k
```

We at StataCorp call files like `test_n_choose_k.do` certification files or, equivalently, certification scripts. Some people might call them validation files. Validation is computer science jargon for providing evidence that software works. Our view is that validation is a process of which certification is a part.

Do certification files really validate results? The first test in `test_n_choose_k.do` asserts that `n_choose_k()` results are equal to $n!/((n-k)!k!)$, as calculated by Mata's `factorial()` function. Under the assumption that Mata's `factorial()` function is validated, and under the assumption that I used the function correctly, I suppose `n_choose_k()` is validated for the value tested. It is validated if you assume that I am right in claiming that $n!/((n-k)!k!)$ is the correct formula for calculating n choose k. Perhaps you would like to know more about me before jumping to that conclusion.

At StataCorp, one part of validation is the process that produces certification files. We obtain valid answers from varied sources. We obtain them by working problems by hand, from textbooks, from other software, and via simulation.

You have just seen the first of what I hope will be many certification files. They are not difficult to write. The one rule is that they stop with error if results are not as expected. We have over 9,000 certification files at StataCorp comprising 1.8 million lines of code. Running them produces 22 million lines of output. It is important to us that certification stop if there is an unexpected result. Flipping through 22 million lines to spot errors would be impossible. Even so, there is another part of validation at StataCorp that involves comparing those 22 million lines with official logs.

There are some rules we do not have at StataCorp: We do not require that certification files be well organized, that they be pretty, or that they be efficient. Our attitude is that the more tests they contain, the better.

9.4   n_choose_k() packaged as an ado-file

Now put out of your mind the do-files we just wrote. We have a different plan for n_choose_k() and its subroutines, which is to write an ado-file to create a new command in Stata. The new command will have the syntax

---

nchooseki # #

When users type **nchooseki 5 2**, for instance, displayed will be the number of ways to choose two things from five:

```
. nchooseki 5 2
  5 choose 2 = 10
```

nchooseki will also store the result in r(comb):

```
. display r(comb)
10
```

We will store the code for this calculator-like command in file nchooseki.ado, as Stata requires. Rather than writing code entirely in Stata, however, the Stata code we write will parse what the user typed to identify the two numbers and call n_choose_k() to make the calculation, and we will put the Mata code for n_choose_k() and its subroutines right in the ado-file with the Stata code.

9.4.1   Writing Stata code to call Mata functions

Here is Stata ado-code that implements nchooseki:

```
────────────────────────── nchooseki.ado ──────────
*! version 1.0.0
program nchooseki, rclass
        version 15
        args n k nothing

        confirm integer number `n´
        confirm integer number `k´
        if ("`nothing´" != "") {
                display as err "`nothing´ found where    ///
                nothing expected"
                exit 198
        }
        mata: st_numscalar("r(result)", n_choose_k(`n´, `k´))
        return scalar comb = r(result)
        display as txt "  `n´ choose `k´ = " as res r(result)
end
Mata code goes here
────────────────────────── nchooseki.ado ──────────
```

The Mata code will appear at the bottom of the ado-file, but right now let's focus on the Stata code and how Stata and Mata communicate with each other. I direct your attention to the highlighted line:

```
mata: st_numscalar("r(result)", n_choose_k(`n', `k'))
```

In Stata, when you code **mata:** ..., Stata invokes Mata to run what follows the colon. If you coded,

```
mata: N = n_choose_k(`n', `k')
```

Stata would invoke Mata, and Mata would execute N = **n_choose_k**('n', 'k'). Stata handles the macro expansion before passing the line to Mata, so if 'n' were 10 and 'k' were 5, Stata would tell Mata to execute the line

```
N = n_choose_k(10, 5)
```

Mata executes the line just as if you had typed it interactively, and the result is the creation of a global Mata variable named N containing 252. Creating a global Mata variable will not help us. We need to get the result back to Stata in the ado-file that we are writing.

Mata provides functions we can use to send results back to Stata:

Save in...	Mata function
Stata global macro	**st_global**(*name*, *stringvalue*)
Stata local macro	**st_local**(*name*, *stringvalue*)
Stata numeric scalar	**st_numscalar**(*name*, *realvalue*)
Stata string scalar	**st_strscalar**(*name*, *stringvalue*)
Stata matrix	**st_matrix**(*name*, *realmatrix*)

The list above is just a subset of what is available. Mata even provides functions to save results in Stata's data. The full set of them is discussed in appendix A. The function we need is **st_numscalar()**. We can use it to save the result of calling **n_choose_k()** by coding

```
mata: st_numscalar("N", n_choose_k(`n', `k'))
```

The above saves the result in a Stata scalar named N, which would solve our problem, but scalar N would be global, and ado-programs should not create globals. We could solve the global issue by coding

```
tempname s
mata: st_numscalar("`s'", n_choose_k(`n', `k'))
```

I, however, am going to use a different solution. Stata commands return results in r(). Mata can return results in r(), too, and **st_numscalar()** will do that if we code "r(*name*)" as **st_numscalar**()'s first argument. We can and will store the result in r(result) by coding

```
mata: st_numscalar("r(result)", n_choose_k(`n´, `k´))
```

That is just what I did. The final lines of the ado-program read

```
mata: st_numscalar("r(result)", n_choose_k(`n´, `k´))
return scalar comb = r(result)
display as txt "  `n´ choose `k´ = " as res r(result)
```

9.4.2 nchooseki.ado

All we have left to do is add the Mata code to the ado-file **nchooseki.ado**. The template for ado-files that contain Mata code is

```
──────────────────────────────── name.ado ────────
*!  version ...                        // see note 1
program name
        version ...                    // see note 2
        ...
end
... more Stata subroutines if necessary ...
                        // Mata portion begins here
version ...                            // see note 3
set matastrict on                      // see note 4
mata:
Mata function, structure, and class definitions
end
──────────────────────────────── name.ado ────────
```

Notes:

1. There are three version statements. This first one is a comment and can be omitted. It is how you track the version of the ado-file. It could be anything you want, such as

   ```
   *! version written 2018.01.24
   ```

2. The second version statement specifies the version of the ado-code interpreter that is to execute the ado portion of the code.

3. The third version statement specifies the version of the Mata compiler that is to be used to compile the Mata functions, structures, and class definitions. Even if this version number is the same as in the second version statement, this version statement must be specified. The version number is not required to be the same as in the second statement, but it usually is.

4. `set matastrict on` is optional but is recommended, even though it will
   be ignored when the ado-file is automatically loaded. It is recommended
   because you can `do` ado-files and then `matastrict` will be switched on.

The **nchooseki** ado-file is

```
────────────────────────────────────────────────── nchooseki.ado ───────────
 1:  *! version 1.0.0
 2:
 3:  program nchooseki, rclass
 4:          version 15
 5:          args n k nothing
 6:
 7:          confirm integer number `n´
 8:          confirm integer number `k´
 9:          if ("`nothing´" != "") {
10:                  display as err "`nothing´ found where nothing expected"
11:                  exit 198
12:          }
13:
14:          mata: st_numscalar("r(result)", n_choose_k(`n´, `k´))
15:          return scalar comb = r(result)
16:          display as txt "  `n´ choose `k´ = " as res r(result)
17:  end
18:
19:
20:  version 15
21:  set matastrict on
22:
23:  mata:
24:
25:  /*        _
26:          N = n_choose_k(n, k)
27:                          -  -
28:
29:                  N = n! / ( (n-k)! k! )
30:
31:          Formula evaluated using canceling to increase domain.
32:  */
33:
34:  real scalar n_choose_k(real scalar n, real scalar k)
35:  {
36:          return( n-k > k ?
37:                  nfactorial_over_kfactorial(n, n-k) /
38:                  nfactorial_over_kfactorial(k, 1)
39:                  :
40:                  nfactorial_over_kfactorial(n, k) /
41:                  nfactorial_over_kfactorial(n-k, 1)
42:                )
43:  }
44:
45:  /*        _____
46:          result = nfactorial_over_kfactorial(n, k)
47:                                              -  -
48:
49:          Returns n!/k! calculated as n*(n-1)*...*(k+1).
50:          Note: 0! = 1.
51:  */
```

```
52:
53:
54:  real scalar nfactorial_over_kfactorial(real scalar n, real scalar k)
55:  {
56:       real scalar     result, i
57:
58:       if (n<0 | n>1.0x+35 | n!=trunc(n))   return(.)
59:       if (k<0 | k>1.0x+35 | k!=trunc(k))   return(.)
60:
61:       result = 1
62:       for (i=n; i>k; --i) result = result*i
63:
64:       return(result)
65:  }
66:  end
```
——————————————————————————————————— nchooseki.ado ——————————

At the top of the Mata code, I included the recommended Stata commands

```
version 15
set matastrict on
```

set matastrict on is ignored when the ado-file is automatically loaded. Mata's strict setting is for your use when you develop the code. Once you release a program into the wild, Mata does not second guess you. If you decide to leave a warning message, so be it. Users will not see it.

So why include the line that Stata and Mata will ignore? You include it because programmers can cause the line not to be ignored. If you were to type do nchooseki.ado, Stata would treat the ado-file as if it were a do-file, meaning that set matastrict on will not be ignored. This means you can debug ado-files in the same way you debug do-files. The proper procedure for running ado-files with do is

```
. clear all           // clear already loaded programs
. do nchooseki.ado    // treat ado-file as do-file and load it
. look at the output
. clear all           // clear so that automatic loading will
                      // not be confused later
```

When you type do nchooseki.ado, the lines of the file will be displayed in the Results window just as they are usually displayed with do-files, and set matastrict on will be executed. Mata will apply the stricter standards as it compiles the code, and you will see any errors or warnings.

I prefer to look at the output in an editor rather than scrolling through the Results window, so I usually type

```
. clear all
. log using check.log, replace
. do nchooseki.ado
. log close
. clear all
. view check.log
```

However you proceed, you should look for warning messages and fix your code so that Mata no longer complains. Treating ado-files as if they are do-files has other advantages as well. Say you are writing `xyz.ado`. I suggest you create a second do-file named `t.do` that contains

```
─────────────────────────────────────────── t.do ───────
clear all
set matastrict on
capture log close
log using t.log, replace
do xyz.ado

tests of xyz go here
log close
─────────────────────────────────────────── t.do ───────
```

You can add tests to `t.do` as you write `xyz.ado`. The tests that I include in `t.do` often test subroutines. When I write ado-code, I open three windows. One is the editor in which I am writing the ado-file, another is the editor in which I am adding tests to `t.do`, and the third is Stata. I can jump from one window to the other, and I can type `do t` whenever I wish.

As development proceeds and more of the code is certified, the tests are deleted or modified to become tests of `xyz` itself. At the end of the development process, the tests in `t.do` become my final certification script. At that point, I delete the line "do `xyz.ado`" and rename the `t.do` file.

9.4.3 test_nchooseki.do

Assume that file `nchooseki.ado` has now been written, and the code even appears to work:

```
. nchooseki 10 5
  10 choose 5 = 252
```

That answer is correct. We need to demonstrate that the code produces other correct answers. Now that we have written `nchooseki.ado`, it should not surprise you that I am going to recommend we write `test_nchooseki.do` to verify that it works.

I created such a file by translating the tests in **test_n_choose_k.do** (from section 9.3.2.2) from Mata to Stata. Mata **assert()** calls became Stata **assert** commands. Here is the file:

```
———————————————————————————————————————————————— test_nchooseki.do ————————
  1:  // version number intentionally omitted
  2:
  3:  clear all
  4:
  5:  // Test 1
  6:  //   For small n, -nchooseki <n> <k>- should produce the same result
  7:  //   as directly calculating n!/((n-k)!k!).
  8:
  9:
 10:  /*
 11:          Certification, test 1.
 12:
 13:          For small n, above code should produce the same result
 14:          as directly calculating n!/((n-k)!k!) for k=0 to n+1.
 15:  */
 16:
 17:  program factorial
 18:          args scaname colon n
 19:          scalar `scaname´ = round(exp(lnfactorial(`n´)), 1)
 20:  end
 21:
 22:  program nk, rclass
 23:          args scaname colon n k
 24:
 25:          tempname nmk nf nmkf kf
 26:
 27:          local nmk = `n´-`k´
 28:
 29:          factorial `nf´    : `n´
 30:          factorial `nmkf´  : `nmk´
 31:          factorial `kf´    : `k´
 32:
 33:          scalar `scaname´ = `nf´/(`nmkf´*`kf´)
 34:  end
 35:
 36:
 37:  program check_small_values
 38:          tempname C
 39:          forvalues n=1(1)5 {
 40:                  local np1 = `n´ + 1
 41:                  forvalues k=0(1)`np1´ {
 42:                          nk    `C´ : `n´ `k´
 43:                          nchooseki  `n´ `k´
 44:                          assert `C´==r(comb)
 45:                  }
 46:          }
 47:  end
 48:  check_small_values
 49:
 50:  nchooseki 5 0
 51:  assert r(comb)==1
 52:  nchooseki 5 6
 53:  assert r(comb)==.
 54:
```

```
55:
56:   // Test 2
57:   //    -nchooseki- should produce results when direct calculation
58:   //    fails.
59:
60:   nchooseki 1000 1
61:   assert r(comb)==1000
62:   nchooseki 1000 2
63:   assert r(comb)==1000*999/2
64:   nchooseki 1000 1000
65:   assert r(comb) == 1
66:   nchooseki 1000 999
67:   assert r(comb) == 1000
68:   nchooseki 1000 998
69:   assert r(comb)==1000*999/2
70:
71:   // Test 3
72:   //    Missing values.
73:
74:   nchooseki -1  1
75:   assert r(comb)==.
76:   nchooseki  1 -1
77:   assert r(comb)==.
78:   rcof "noisily nchooseki  5   ." == 7
79:   rcof "noisily nchooseki  .  2" == 7
80:   rcof "noisily nchooseki  .  2" == 7
81:   rcof "noisily nchooseki  .   ." == 7
82:   rcof "noisily nchooseki  .  2" == 7
83:
84:   // Test 4
85:   //    Invalid syntax.
86:
87:   rcof "noisily nchooseki"          == 7
88:   rcof "noisily nchooseki 10"       == 7
89:   rcof "noisily nchooseki 10 5 3" == 198
```
——————————————————————————————— test_nchooseki.do ———————

Test 4 at the bottom of the file is new. It verifies that **nchooseki** responds appropriately when too few or too many arguments are specified. Stata's **rcof** command is a variation on Stata's **assert** command. **rcof** verifies that the return code from running the quoted command is as stated. To learn more, type **help rcof**.

All I need to do to test **nchooseki.ado** is type

```
. do test_nchooseki
(output omitted)

. _
```

9.4.4   Mata code inside of ado-files is private

The Mata code that appears inside ado-files is private to the ado-file. `nchooseki.ado`
contains Mata functions `n_choose_k()` and `nfactorial_over_kfactorial()`. These
functions can be used only by code that is inside the ado-file. Outside of the file, the
functions simply do not exist.

```
. nchooseki 10 2
  10 choose 2 = 45
. mata: n_choose_k(10, 2)
                <istmt>:  3499  n_choose_k() not found
  r(3499);
```

There is an advantage and a disadvantage to privacy. The advantage is that you can
name functions as you please. The disadvantage is that testing is more difficult. Because
`n_choose_k()` does not exist outside the ado-file, you cannot test it directly. I worked
around the inability to test `n_choose_k()` by thoroughly testing `nchooseki`.

You might be tempted to use the tricks I have shown you to test `n_choose_k()` directly.
I appreciate the temptation, but as a general rule, certification files should not exploit
insider information on how the command is written. You might need to update the
program one day, and it is useful if old test scripts can be used to test updated code.
Notice that the certification script in the previous section did not include the line

```
. do nchooseki.ado
```

`test_nchooseki.do` lets the ado-file load automatically. Final testing should be done
in the environment in which users will be running your code.

During the development process, however, it is useful to reach inside and test subroutines, or at least try them interactively, and it is useful to do that whether the
subroutines are written in ado or Mata. It is useful to treat the ado-file as a do-file:

```
. clear all
. do nchooseki.ado
```

Type the above, and all the routines inside it are publicly exposed and directly accessible.

9.5   n_choose_k() packaged as a Mata library routine

A Mata library is a file containing Mata functions, structures, and classes. The file
does not contain the original source code; it contains the ready-to-execute compiled
code. Once a function, structure, or class is stored in a library, it becomes part of
Mata. If `n_choose_k()` were in a library, we could use the function in any context, be
it interactively, in do-files, in ado-files, or in writing other Mata functions.

Libraries are easily shared because each is a single file.

To put Mata code in a library, first develop the code packaged as a do-file. We created two do-files when we did that in section 9.3:

$$\begin{array}{ll}\texttt{n\_choose\_k.mata} & \text{source code} \\ \texttt{test\_n\_choose\_k.do} & \text{certification file}\end{array}$$

These files are just what we need. We discussed the creation of libraries in section 2.4. We put the code from file `hello.mata` into a library by running a do-file, `make_lmatabook.do`, which was

```
─────────────────────────────── make_lmatabook.do ───────────
// version number intentionally omitted

clear all
do hello.mata

lmbuild lmatabook.mlib, replace
─────────────────────────────── make_lmatabook.do ───────────
```

All we do is add the line "`do n_choose_k.mata`" to `make_lmatabook.do`:

```
─────────────────────────────── make_lmatabook.do ───────────
// version number intentionally omitted

clear all
do hello.mata
do n_choose_k.mata

lmbuild lmatabook.mlib, replace
─────────────────────────────── make_lmatabook.do ───────────
```

Then we type

```
. do make_lmatabook
```

I have more to say, but before we continue with this, let's stop to consider where we should put the files `make_lmatabook.do`, `n_choose_k.mata`, and `test_n_choose_k.do`.

9.5.1 Your approved source directory

You need to create a directory where you will keep the files related to the functions you place in libraries. Where you locate it and how you name it is up to you. We will just call it your Approved Source Directory, or ASD for short. You will need to treat the directory's contents with caution.

You will not, for instance, develop code in your ASD. You will copy code into your ASD from wherever it was that you developed it, and you will copy it only after you are reasonably convinced that the code is perfect. You will be wrong about that sometimes, which brings me to my second point.

You will keep backups of your ASD. This way, if you make a mistake, you can restore the directory to how it was previously. The danger is not copying a new file to the directory—after all, you could just erase the file. The danger is when you replace existing files in your ASD because they need updating.

Let's imagine your ASD already exists and has important files in it, among them being n_choose_k.mata and test_n_choose_k.do. You decide to make an improvement to n_choose_k.mata. The right way to proceed is to copy the n_choose_k.mata and test_n_choose_k.do somewhere else to work on them. When you are reasonably convinced that you have made them better, then you copy the files back. It may happen that you later realize the new files are seriously inferior to what you previously had. If you have backed up the directory, you can restore the original files.

lmatabook.mlib will be the library where we will put the compiled code developed in this book. You can use it for your own code, too, or you can create other libraries. Either way, you will use the same ASD. To add a second library, you create another make_l*name*.do file. All your .mata files go in your ASD regardless of which library contains the compiled code.

Your ASD will comprise the files listed in table 9.1:

Table 9.1. Your approved source directory (ASD)

Filename	Contents
make_lmatabook.do	Do-file to create or re-create Mata library lmatabook.mlib
test.do	Do-file to run all certification files
hello.mata	Source code for hello() and goodbye()
n_choose_k.mata	Source code for n_choose_k() and subroutine
test_n_choose_k.do	Certification file for n_choose_k.mata

Notes:
1. lmatabook.mlib is the name of the library where we will put the compiled code developed in this book. You can create other libraries in the future, but right now, we will pretend that lmatabook.mlib is your only library.
2. Mata library names start with the letter l (a lowercase L) and end in .mlib. That is why the library for this text, *The Mata Book*, is named lmatabook.mlib and not matabook.mlib.
3. lmatabook.mlib (and other libraries you may create) are not stored in your ASD. Libraries are stored in your PERSONAL directory. You can type sysdir at the Stata prompt to find out where your PERSONAL directory is.
4. Typing do make_lmatabook will create or re-create the lmatabook.mlib library and store it in your PERSONAL directory. It will even create your PERSONAL directory if the directory does not already exist.
5. Templates for make_lmatabook.do and test.do are provided below, including templates for making additional libraries.

Whether you have one library or many, you can put .mata files in your ASD that appear in no library, too. Your ASD is a good place to store all final code. This way, if you ever need to find a particular .mata file, you will not need to wonder where the file is.

Your test_*name*.do files are also placed in your ASD. File test.do will run all the test_*name*.do files. You store the certification files together and you run them together because the files get better at detecting bugs the more files there are. They get better because code is interdependent. Imagine that you wrote a function called allperms() that calls n_choose_k(). You therefore add files allperms.mata and test_allperms.do to your ASD. File test_allperms.do may be intended to test allperms(), but it indirectly tests n_choose_k(), too. Years from now, you might modify n_choose_k(), and test_allperms.do might reveal a new error in n_choose_k() that test_n_choose_k.do does not.

Below, I describe each of the files stored in the ASD.

9.5.1.1   make_lmatabook.do

File make_lmatabook.do creates or re-creates the Mata library lmatabook.mlib. The template is

```
─────────────────────────────────── make_lmatabook.do ──────────
// version number intentionally omitted
clear all
do hello.mata
do n_choose_k.mata
lmbuild lmatabook.mlib, replace
─────────────────────────────────── make_lmatabook.do ──────────
```

As this book proceeds, we will be adding additional .mata files to our ASD and adding do *name*.mata lines to make_lmatabook.do.

You can create other make_l*name*.do files to create other libraries. If you want to store some of your code in library lsmith.mlib, then create do-file make_lsmith.do:

```
─────────────────────────────────── make_lsmith.do ──────────
// version number intentionally omitted
clear all
do ...
do ...
    .
    .
lmbuild lsmith.mlib, replace
─────────────────────────────────── make_lsmith.do ──────────
```

Remember that Mata library names must start with the letter l (a lowercase L).

9.5.1.2 test.do

The template for `test.do` is

```
———————————————————————————— test.do ————————
// version number intentionally omitted
do test_n_choose_k
———————————————————————————— test.do ————————
```

As the book proceeds, we will add other **do** test_*name* lines to the list.

As you write your own code, you add additional **test_***name***.do** files to your ASD and additional **do** **test_***name* lines to **test.do** regardless of the library in which the compiled code will be stored or if it is even stored in a library at all.

9.5.1.3 hello.mata

File `hello.mata` is from chapter 2. It is

```
———————————————————————————— hello.mata ————————
 1:  version 15
 2:  set matastrict on
 3:
 4:  mata:
 5:  void hello()
 6:  {
 7:          printf("hello, world\n")
 8:  }
 9:
10:  void goodbye()
11:  {
12:          printf("good-bye, world\n")
13:  }
14:  end
———————————————————————————— hello.mata ————————
```

9.5.1.4 n_choose_k.mata

File `n_choose_k.mata` is from section 9.3.2.1. It looks like this:

```
──────────────────────────────────────── n_choose_k.mata ────────
version 15
set matastrict on
mata:
real scalar n_choose_k(real scalar n, real scalar k)
{
            .
            .
}
real scalar nfactorial_over_kfactorial(real scalar n,
                                       real scalar k)
{
          .
          .
}
end
                                        ──────── n_choose_k.mata ────────
```

The entire file can be found in section 9.3.2.1.

9.5.1.5 test_n_choose_k.do

File `test_n_choose_k.do` is from section 9.3.2.2. It looks like this:

```
──────────────────────────────────────── test_n_choose_k.do ────────
// version number intentionally omitted
clear all
run n_choose_k.mata
mata:
   .
   .
end
                                        ──────── test_n_choose_k.do ────────
```

The entire file can be found in section 9.3.2.2.

9.5.2 Building and rebuilding libraries

To build or rebuild the `lmatabook.mlib` library, in Stata, type

```
. cd ASD              // change to your ASD
. do test             // verify files pass certification
  (output omitted)
```

```
. do make_lmatabook    // build/rebuild library
  (output omitted)
. do test              // verify files pass certification again
  (output omitted)
. cd someplace else     // change away from your ASD
```

If you wanted to rebuild `lsmith.mlib`, you would substitute do `make_lsmith` for do `make_lmatabook`.

Notice that `test.do` is run twice, once before building the library and then again afterward. That is important. It is the second run that tests interdependencies. You run `test.do` the first time to ensure that there are no apparent problems. After rebuilding the library, you run it again to prove that there are no problems. Imagine that you typed the above because you updated `n_choose_k.mata`. The first run of `test.do` establishes that the code in `n_choose_k.mata` passes the tests in `test_n_choose_k.do`. The other certification files continue to use the original `n_choose_k()` function stored in the library. The second run of `test.do` differs from the first because the library has been updated. All the certification files are now using the updated `n_choose_k()` function, and if the functions they test call `n_choose_k()`, they might detect a bug in it.

9.5.3   Deleting libraries

Deleting a library is safe to do because, if you delete it mistakenly, you just have to type do `make_libname` to re-create it.

To delete `lmatabook.mlib`, type

```
. erase "`c_sysdir_personal´/lmatabook.mlib"
. mata: mata mlib index
```

The last line causes Mata to rebuild its library index.

10 Mata's structures

10.1 Overview

Structures are programming constructs that make it easier to deal with lots of variables. I will show you an example of programming with structures in the next chapter. In this chapter, I define structures and tell you about them.

Saying that structures make it easier to deal with lots of variables understates their importance. Structures allow you to avoid writing code that looks like this:

```
void obtainb(real matrix X,  real colvector y,
             real scalar include_intercept,
             real colvector b, real matrix V,
             real scalar R2,
             real scalar F, real scalar ndf, real scalar ddf,
             real scalar K, real scalar K_adj, real scalar k,
             real scalar N)
{
        .
        .
        .
}
```

Function `obtainb()` has too many arguments. There may be too many, but there is no getting rid of them because each serves a purpose. Structures get around this problem by providing variables that themselves contain other variables. The structure variable might be named `lr`, and the variables it contains might be `lr.X`, `lr.y`, and all the other variables listed above. Using structures, function `obtainb()` might have just one argument:

```
void obtainb(struct Lr scalar lr)
{
        .
        .
        .
}
```

Inside the body of the function, the program would now refer to `lr.X`, `lr.y`, `lr.include_intercept`, and so on.

I am imagining that function `obtainb()` is intended to calculate linear regression coefficients. In its original form, the first 3 arguments—`X`, `y`, and `include_intercept`—were the inputs, and the remaining 10 were for receiving the results that the function calculates.

The one-argument form of `obtainb` is better, but it would be better still if we divided the 13 arguments into inputs and outputs. We could create structure variable `inputs` to contain `inputs.X`, `inputs.y`, and `inputs.include_intercept`, and we could create structure variable `results` to contain the 10 outputs. The function would then be

```
/*
    void obtainb(inputs,  results)
                 ------
*/

void obtainb(struct LrInputs  scalar inputs,
             struct LrOutputs scalar results)
{
        .
        .
        .
}
```

At this point, it would be even more elegant to have the function simply return **results** instead of specifying **results** as an argument, which we can do by coding

```
/*      _____
        results = obtainb(inputs)
                          ------
*/

struct LrOutputs scalar obtainb(struct LrInputs scalar inputs)
{
        struct LrOutputs scalar   results
        .
        .
        .
        return(results)
}
```

We have come a long way from a function with 13 arguments. The one-argument code might look like this:

```
struct LrOutputs scalar obtainb(struct LrInputs scalar inputs)
{
        struct LrOutputs scalar   results
        .
        .

        if (!inputs.include_intercept) {
                XX = inputs.X'inputs.X
                Xy = inputs.X'inputs.y
        }
        else {
                XX = ...
                Xy = ...
        }
        results.b = invsym(XX)*Xy
        results.V = ...
        .
        .
        .
        return(results)
}
```

The result is more readable code. The code for calling `obtainb()` will be equally readable:

```
void run_regression(real colvector y, real matrix X,
                    real scalar include_intercept)
{
        struct LrInputs  scalar    i
        struct LrOutputs scalar    r

        i.X                 = X
        i.y                 = y
        i.include_intercept = include_intercept

        r                   = obtainb(lr_in)

        // code to display r.b, r.V, ..., r.N appears here
}
```

10.2   You must define structures before using them

Before you can use variables like `inputs` and `results`, you must define the variables that they are to contain. You do that with structure definitions. Here are the definitions of `LrInputs` and `LrOutputs`:

```
struct LrInputs
{
        real matrix    X
        real colvector y
        real scalar    include_intercept
}
struct LrOutputs
{
        real colvector b
        real matrix    V
        real scalar    R2
        real scalar    F, ndf, ddf
        real scalar    K, K_adj, k, N
}
```

You define `LrInputs` and `LrOutputs` outside the functions that use them just as you define subroutines outside the functions that call them. Once `LrInputs` and `LrOutputs` are defined, you can declare `struct LrInputs` and `struct LrOutputs` variables just as you can declare `real`, `string`, and other types of variables:

```
struct LrOutputs scalar obtainb(struct LrInputs scalar inputs)
{
        struct LrOutputs scalar    results
        .
        .
        .
        return(result)
}
```

10.3 Structure jargon

If we were discussing integers and not structures, I would start by telling you the definition of integers. They are whole numbers. Then I would offer some examples and tell you that instances of integers include 1, 2, 3, and so on.

So it is with structures. The *definition* of `LrOutputs` is

```
struct LrOutputs
{
        real colvector b
        real matrix    V
        real scalar    R2
        real scalar    F, ndf, ddf
        real scalar    K, K_adj, k, N
}
```

If we declare

```
struct LrOutputs scalar    r
```

we made **r** an *instance* of structure `LrOutputs`.

In the computer science literature, the word *structure* is invariably followed by the word *definition* or *instance* as if you might confuse one with the other if it were omitted. If **r** equals

```
r.b          = (-0.0060, 0.0005)
r.V          = ( 2.6782e-7, -0.0008 \ -0.0008, 2.6050)
r.R2         = 0.6529
r.F          = 66.79
r.ndf        = 2
r.ddf        = 72
  etc.
```

I somehow feel obligated to write that **r** is an instance of structure `LrOutputs` even though, if I were just talking to you, I would probably just say that **r** is an `LrOutputs`, just as I would say that 5 is an integer when I should say that 5 is an instance of an integer.

I mention this because I do not want you to puzzle over the deep meaning of "instance" as I did when I first learned about structures. Just as there is one definition for integers and many instances of them, in a program, there is one definition of a structure and (perhaps) many instances. Just as I can have three different **real scalar** variables by declaring

```
real scalar    a, b, c
```

I can have three different `LrOutputs` variables by declaring

```
struct LrOutputs scalar    r, s, t
```

10.4   Adding variables to structures

Use of structures make code easier to modify. Months after writing `obtainb()`, you might need to add the calculation of t statistics to the code. If so, all you need to do is add `real colvector t` to the definition of `LrOutputs` and add code for calculating `results.t` to `obtainb()`.

Without structures, you would have needed to add a fourteenth argument to `obtainb()`, and users would have to modify and recompile their code even if they had no use for the new statistics. With structures, they need only recompile, and if you put the new variable at the bottom of the structure, they do not even have to do that.

10.5   Structures containing other structures

Structures can contain other structures, such as

```
struct LrProblemAndSolution
{
        string            scalar     problem_name
        struct LrInputs   scalar     inputs
        struct LrOutputs  scalar     results
}
```

This structure has either 3 members or 16 depending on whether you count the members' members. Let variable `ps` be an **LrProblemAndSolution**. Obviously, `ps` at least has members `ps.problem_name`, `ps.inputs`, and `ps.results`.

`ps.inputs` and `ps.results` are themselves structures, however, and you can access their members by coding `ps.inputs.X`, `ps.inputs.y`, `ps.inputs.include_intercept`, and so on.

A program that uses **LrProblemAndSolution** might read

```
void run_regression(real colvector y, real matrix y,
                    real scalar include_intercept)
{
        struct LrProblemAndSolution scalar    ps

        ps.inputs.X                   = X
        ps.inputs.y                   = y
        ps.inputs.include_intercept   = include_intercept

        ps.results                    = obtainb(ps.inputs)

        // code to display
        //    ps.results.b, ps.results.V, ..., ps.results.N
        // appears here
}
```

10.6   Surprising things you can do with structures

Structure variables are variables, and therefore you can pass and return them just like you can pass and return other variables. You can also assign values from one structure to another and test whether two structures are equal. When you code

```
s2 = s1
```

the contents of s1 are copied to s2. That includes s2's members, s2's members' members, and so on. You can test whether structures are equal by coding

```
if (s1 == s2) ...
```

You can test whether they are not equal by coding

```
if (s1 != s2) ...
```

Two structures are equal if 1) they are both structures, 2) they are of the same type (say, LrProblemAndSolution), and 3) their members are equal, as well as their members' members, and their members' members' members, and so on. If you coded

```
if (s1 == 2) ...
```

the condition would be false. It is not an error.

10.7   Do not omit the word scalar in structure declarations

The most common error that even experienced programmers make is to omit the word scalar in structure declarations, as I will now do in the declaration of variable in:

```
void run_regression(real colvector y, real matrix y,
                    real scalar include_intercept)
{
        struct LrInputs        in          // <- scalar missing
        in.X = ...
             .
             .
             .
}
```

The first warning of the problem will come when you compile the code, but only if you have matastrict set on. Mata will report "note: variable in may be used before set". The message is admittedly not clear about what the problem might be. If you execute the code anyway, Mata will abort execution of the code and issue an error:

```
: run_regression(y, x, 1)
      run-regression():  3259  nonstruct found where struct
> required
                <istmt>:     -  function returned error
r(3259);
```

This message, too, is not clear about what the problem is, so let me disentangle it for you. The error message is referring to the line

```
in.X = ...
```

The message is saying that `in` is not a structure. How can that be? The earlier declaration states emphatically that `in` is indeed a structure:

```
struct LrInputs      in
```

The problem is that I omitted orgtype `scalar` from the declaration, which resulted in Mata compiling the code as if I had specified the default orgtype, `matrix`:

```
struct LrInputs matrix      in
```

This resulted in variable `in` being initialized as a 0×0 matrix of structures, as explained in section 6.2.1. After declaration, matrices are 0×0. Thus, variable `in` is 0×0, and so in the assignment `in.X = ...`, there is no `in.X` to fill in. `in` itself barely exists. Had I coded orgtype `scalar`, `in` would have been 1×1 and then there would have been an `in.X` to which a value could be assigned. But I did not do that. I omitted the word `scalar`, so `matrix` was assumed, and 0×0 variable `in` has no member `in.X`.

The fix in this case is to go back to the declaration and insert the word `scalar`.

At least, that is the solution unless you really wanted to create a matrix of structures. How you do that is the topic of the next section.

10.8   Structure vectors and matrices and use of the constructor function

You may create vectors and matrices of structures. Assume you need to store red-green-blue color values corresponding to objects that you will display. A good way to store the colors would be in a structure, such as

```
struct Color {
        real scalar   red
        real scalar   green
        real scalar   blue
}
```

In this case, you might indeed want a vector of colors, or even a matrix of them. In the following code, after `c` is declared to be a `Color` vector, `c` will be 1×0.

```
... foo(...)
{
        struct Color vector  c
        .
        .
        .
}
```

Before you can use c, you must initialize it to contain the desired number of structures.

When you declare a structure, Mata automatically creates a function of the same name that creates structure instances. For example, the declaration of structure Color automatically created function Color(). Function Color() is called Color's constructor function.

Color() allows zero, one, or two arguments.

Color() without arguments returns a 1×1 instance of **struct Color**. In the preceding section, where variable in was declared to be an **LrInputs matrix** because I forgot to specify **scalar**, the code would have worked if I had coded in = LrInputs() before assigning a value to in.X. Before the assignment, in was 0×0. After the assignment, it would have been 1×1.

Color(n) returns n new instances of **struct Color** arranged as a $1 \times n$ vector.

Color(r, c) returns $r \times c$ new instances of **struct Color** arranged as an $r \times c$ matrix.

The following code returns a color matrix of the same dimension as **object** and sets the color to white:

```
struct Color matrix default_color_of_object(transmorphic object)
{
        struct Color matrix  c
        real scalar          i, j
        c = Color( rows(object), cols(object) )
        for (i=1; i<=rows(c); i++) {
                for (j=1; j<=cols(c); j++) {
                        c[i,j].red   = 255
                        c[i,j].green = 255
                        c[i,j].blue  = 255
                }
        }
        return(c)
}
```

The following function does the same thing, but more elegantly:

```
struct Color matrix default_color_of_object(transmorphic object)
{
        struct Color matrix   c
        struct Color scalar   white
        real scalar           i, j

        white.red = white.green = white.blue = 255

        c = Color( rows(object), cols(object) )

        for (i=1; i<=rows(c); i++) {
                for (j=1; j<=cols(c); j++) c[i,j] = white
        }
        return(c)
}
```

Notice how members of structures are accessed when dealing with a matrix of structures. `c[i,j].red` is how you refer to the (i, j) structure's `red` value. You do not refer to it as `c.red[i,j]`. That would refer to a scalar structure's value of `red[i,j]`, meaning that `red` would be the matrix, not `c`.

10.9   Use of transmorphic with structures

Section 6.3.2.1 discussed the use of `transmorphic` for declaring passthru variables. A variable is passthru when the function does not directly access its contents. Variable `db` is passthru in the following example:

```
string scalar getinfo(...)
{
        .
        .
        struct DbStatus scalar   db
        db = open_database(...)
        .
        .
        X = extract_from_database(db, x, ...)
        .
        .
        close_database(db)
        return(...)
}
```

Variable **db** contains the information that the database functions need to access the data. Structure **DbStatus** might be defined as

```
struct DbStatus
{
        string scalar  db_name
        real scalar    db_handle
        real scalar    db_status, db_location
        string scalar  db_request
}
```

The database functions that **getinfo()** called would look like this:

```
struct DbStatus scalar open_database(...)
{
        struct DbStatus scalar    db
        .
        .
        return(db)
}
string scalar extract_from_database(
                        struct DbStatus scalar db,
                        ...)
{
        string scalar    s
        .
        .
        return(s)
}
void close_database(struct DbStatus scalar db)
{
        .
        .
}
```

Nonetheless, declaring **db** as a structure was unnecessary in **getinfo()** because that function never looked inside **db**. **getinfo()** obtained **db** from **open_database()** and passed **db** to the other database functions. Even though **db** is in fact a **struct DbStatus**, it would have been better to declare it **transmorphic**:

```
string scalar getinfo(...)
{
        .
        .
    transmorphic db
    db = open_database(...)
        .
        .
    X = extract_from_database(db, ...)
        .
        .
    close_database(db)
    return(...)
}
```

Declaring db as **transmorphic** is the same as declaring it as **transmorphic matrix**. That means db starts out as being 0×0, but that is okay because db is reassigned by the line

```
db = open_database(...)
```

Declaring db as **transmorphic** ensures that getinfo() cannot access db's member variables, which means that getinfo() will not need to be recompiled if the structure definition is updated.

10.10 Structure pointers

Pointers can be used with structures. Let s be a **struct S scalar** with members a and b, and assume p = &s. Then *p is a synonym for s just as it would be if s were not a structure. I mentioned in previous chapters that you can type *p any place you can type s, but that is not true with structure pointers. If you want to type the synonyms for s.a and s.b, you have to type (*p).a and (*p).b. You have to include the parentheses because the dot separator has higher precedence than the * operator. If you typed *p.a, that would be interpreted as *(p.a), which means something entirely different. It means the contents of the pointer p.a, meaning that p would be a structure and a would be a pointer.

If you find this confusing, you are not alone. Because of that, an alternative syntax is provided for dealing with pointers to structures. When p = &s, synonyms for s.a and s.b can be written as p->a and p->b, which are read aloud as "p pointing to a" and "p pointing to b". Using the -> operator makes code more readable.

Consider the following structure definition:

```
struct S
{
        real scalar                             a, b
        struct S2 scalar                        c
        pointer(struct S3  scalar) scalar  q
}
```

This structure is just like the one I described with members a and b, and with two extra members, c and q. Member c is itself another structure, S2, and member q is a pointer to yet another structure, S3.

If p = &s, then

If you want...	You code...	Meaning
s	*p	entire S
s.a	p->a	member a of S
s.b	p->b	member b of S
s.c	p->c	entire S2
s.c.mbr	p->c.mbr	member mbr of S2
s.q	p->q	pointer to entire S3
*s.q	*p->q	entire S3
*s.q->mbr	p->q->mbr	member mbr of S3
&s	p	pointer to entire S
&s.a	&(p->a)	pointer to a of S
&s.b	&(p->b)	pointer to b of S

Note: p->a and p->b can be coded (*p).a and (*p).b, but no one does that.

Pointers to structures are not used as often in Mata as they are in C, C++, and some other languages, because some problems that require their use in those languages do not require their use in Mata. For instance, consider the following Mata code:

```
void setup(struct simple scalar s) { s.a = s.b = 0
}

void example()
{
        struct Simple scalar   t
        setup(t)
        .
        .
}
```

When function example() calls setup(t), it will result in t.a and t.b being set to 0.

In some languages other than Mata, calling `setup(t)` would not result in `t.a` and `t.b` being changed, because those languages pass copies of arguments. Mata does not work that way. In the other languages, you would write `setup()` differently, using pointers. Programmers of the other languages would write the routines like this:

```
void setup2(pointer(struct simple scalar) scalar p)
{
        p->a = p->b = 0
}
void example2()
{
        struct Simple scalar  t
        setup2(&t)
        .
        .
}
```

In Mata, `setup()` and `setup2()` each work equally well. You should code in the way you find most natural. If you find neither natural yet, then code in the style of `setup()`. As far as Mata is concerned, `setup2()` introduces needless complication.

Pointers nonetheless have their uses. One is to conserve memory. At the top of this chapter, we had the structure `LrInputs`. It was

```
struct LrInputs
{
        real matrix     X
        real colvector  y
        real scalar     include_intercept
}
```

We were discussing a linear-regression subroutine named `obtainb()`. A problem with the `obtainb()` code that I did not mention is that it wastes memory. The waste arises because callers of `obtainb()` will already have an X matrix and a y vector, and I required callers to put a copy of X and y in the structure before they could call `obtainb()`. To remind you, callers of `obtainb()` were required to write code that looked like this:

```
inputs.X = X
inputs.y = y
inputs.include_intercept = 1
outputs = obtainb(inputs)
```

What I required is an elegant design for functions with lots of input and output variables, but it was wasteful of memory. We can keep the elegant design and eliminate the waste if we change **LrInputs** to use pointers:

```
struct LrInputs
{
        pointer(real matrix)     scalar    *X
        pointer(real colvector)  scalar    *y
        real scalar                        include_intercept
}
```

The advantage of using pointers is that **inputs.X** and **inputs.y** will each require only eight bytes of memory because that is how much memory it takes to record the address of an object regardless of the object's size.

Using this new **LrInputs**, we will now require callers of **obtainb()** to set **inputs.X** and **inputs.y** using the address-of operator:

```
inputs.X = &X
inputs.y = &y
inputs.include_intercept = 1
outputs = obtainb(inputs)
```

The pointer method for holding onto objects while conserving memory appears over and over in programs. It can be used whenever inputs might be large, as vectors, matrices, and even structures can be.

11 Programming example: Linear regression

11.1 Introduction

We are going to use structures in this chapter to implement code to perform linear regression. I am not thinking about a textbookish, oversimplified example. I am thinking about a routine that mimics Stata's **regress** command, a routine so good that it could become an official part of Mata. Achieving that goal would take the rest of this book, so we will not go the full distance, but we will go farther than you might expect, and it will be obvious how you could finish what we have started.

Linear regression concerns the solving of formulas such as $\mathbf{y} = \mathbf{Xb}$ for $\mathbf{b}$. As Stata users know, there is more to linear regression than just solving for $\mathbf{b}$. Should the model include an intercept? Do you want to know the variance matrix associated with **b**? Do you want the predicted values of **y** based on **b**? The residuals? I could continue.

An obvious way to arrange the code would be to define a structure with variables to be filled in with 0s or 1s, in which users could specify answers to each of the above questions. It might look like this:

```
struct R_options
{
        real scalar    addcons    // 0,1: whether intercept added
        real scalar    V          // 0,1: whether VCE calculated
        real scalar    yhat       // 0,1: whether yhat calculated
        real scalar    resids     // 0,1: whether residuals
                                  //            calculated
        .
        .
        .
}
```

We might define another structure where calculated results could be stored:

```
struct R_results
{
        real colvector    b
        real matrix       V
        .
        .
        .
        real colvector    yhat
        real colvector    resids
}
```

With these two structures, the code could look something like this:

```
struct R_results scalar regress(real colvector  y,
                                real     matrix  X,
                     struct R_options   scalar  options)
{
        struct R_results scalar   res
        .
        .
        .
        res.b = ...
        if (options.V) {
                res.V = ...
        }
        if (options.yhat) {
                res.yhat = ...
        }
        .
        .
        .
        return(res)
}
```

Users would code calls to **regress()** like this:

```
options.addcons          = 1
options.vce              = 1
options.predicted_values = 0
options.resids           = 0
res = regress(y, X, options)
```

The above is the obvious design, and it mirrors the design of Stata's **regress** command, but it is not the design we will be using. We are instead going to write code in a style known as self-threading code. I will show you what the code will look like, but the details of the code will make no sense to you yet. All that you need to notice is the proliferation of subroutines, such as **lrset()**, **lr_b()**, **lr_V()**, **lr_yhat()**, and **lr_resids()**.

```
struct lrinfo scalar lrset(real colvector  y,
                           real     matrix  X,
                           real     scalar  addcons)
{
        r.y       = &y
        r.X       = &X
        r.addcons = addcons

        r.b = r.V = ... = .z
        return(r)
}
real colvector lr_b(struct lrinfo scalar lr)
{
        if (r.b == .z) {
                r.b = ...
        }
        return(r.b)
}
real matrix lr_V(struct lrinfo scalar lr)
{
        if (r.V == .z) {
                r.V = ...
        }
        return(r.V)
}
real colvector lr_yhat(struct lrinfo scalar lr)
{
              .
              .
}
real colvector lr_resids(struct lrinfo scalar lr)
{
              .
              .
}
```

The `lr_*()` subroutines take the place of **struct R_options** in the previous code. Users of the system—programmers like us—will call the subroutines `lr_b()`, `lr_V()`, ..., `lr_yhat()`, and `lr_resids()`, or not call them, depending on whether they want the particular calculated result.

You write programs for use by Mata users differently from how you would write them for use by Stata users. If you were writing for Stata users, you might indeed write a **regress()** function with all of its options. When you write code in Mata, however, you are writing for other Mata users, and they want tools for building other programs. These tools, if they are well written, would make it easy to write a **regress** command. It would be easy to implement

```
struct R_results scalar regress(real colvector   y,
                                real     matrix   X,
                struct R_options       scalar   options)
{
        struct R_results scalar    res
        struct lrinfo scalar       r

        r = lrset(y, X, options.addcons)

        res.b = lr_b(r)
        if (options.V)        res.V      = lr_V(r)
        .

        .
        if (options.yhat)    res.yhat   = lr_yhat(r)
        if (options.resids)  res.resids = lr_resids(r)

        return(res)
}
```

The **regress()** function would be easy to write, but I doubt any Mata user would bother to write it or even use it if it were written for them because the `lr_*()` functions themselves are so easy to use.

This system of subroutines has another advantage, too. Because it has one subroutine for each calculated result, the code will be easier to maintain and extend in the future. If we want to change a result, we modify its subroutine. If we want to add a new result, we add a new subroutine.

It is not possible to exaggerate the importance of writing modifiable and extensible code. All the good design in the world will not circumvent the need to modify the code and even extend its design later. As an example, there are few statistical calculations more settled than that of linear regression, yet we at StataCorp have had to modify Stata's regression command many times. Here is an abbreviated history:

1. **regress** was implemented.

2. **regress** was modified to include instrumental variables.

3. **regress** was modified to be more accurate with larger datasets because desktop computers became more powerful and had more memory.

4. **regress** was modified to include robust standard errors.

5. **regress** was modified to include population weights.

6. **regress** was modified to include clustering.

7. **regress** was modified to be more accurate yet again because desktop computers became even more powerful and the datasets analyzed grew all the more.

8. **regress** was modified to support parallel computing.

9. **regress** was modified to better support parallel computing as chips used improved designs to speed communication between processors.

10. **regress** was modified to support factor-variable notation, which required a complete change in how **regress** handled multicollinearity.

Most of the modifications could not have been predicted. Serious programmers write code in a style that can be easily modified, and the ultimate in modifiability is self-threading code.

11.2 Self-threading code

Self-threading code is easiest understood by looking at an example. Consider the following interrelated calculations:

For given values of a and b, the values of x, y, and z are given by

$$x = (a + b) * z$$
$$y = 2 * b * x$$
$$z = a - b$$

These formulas are stand-ins for the more complicated formulas we will be programming later. These formulas have the virtue of being simple while still preserving the complication that some calculated results depend on other calculated results. Whether you care about z or not, you must calculate it before x or y can be calculated, and you also must calculate x before y can be calculated.

Self-threading code can calculate, for any number of related formulas, all or any subset of them, make only the minimum number of calculations required, and do that without knowing which calculations will be requested, the order in which they will be requested, or even the number that will be requested. Later in this chapter, we will use self-threading code to implement the 15 formulas for calculating linear regression, but first, let's understand how self-threading code works with our 3-formula example.

Variables a and b in the example are called the input variables.

Variables x, y, and z are called the derived variables.

First, we define a structure to contain the input and derived variables:

```
struct info
{
        // ----------------------- input variables ---
        real scalar    a
        real scalar    b

        // --------------------- derived variables ---
        real scalar    x
        real scalar    y
        real scalar    z
}
```

Next, we write separate functions to calculate each derived variable. I will name the functions c_x(), c_y(), and c_z(). Here is c_x(), the routine for calculating x:

```
real scalar c_x(struct results scalar r)
{
        if (r.x has not yet been calculated) {
            r.x =   (r.a+r.b)*c_z(r)
            mark r.x as calculated
        }
        return(r.x)
}
```

The code is deceptively simple. If x has been previously calculated, then c_x() returns it. Otherwise, c_x() calculates x and stores the value in r.x so that it can be returned now and again in the future if the need should arise.

Notice that in making the calculation for x, c_x() needs z. c_x() obtains z by calling c_z(). c_z() will be written in the same style as c_x(). If z has already been calculated, then the value is returned. If it has not yet been calculated, then the value is calculated, stored, and returned.

How does the code determine whether values have already been calculated? In Mata, the easy solution is to initialize each derived variable with a special value and code the corresponding calculation routine to check for that value. In the code below, I will use the value .m, one of Mata's missing-value codes. The final code for c_x() is then

```
real scalar c_x(struct results scalar r)
{
        if (r.x == .m) {
            r.x =   (r.a+r.b)*c_z(r)
        }
        return(r.x)
}
```

c_x() uses .m and not the ordinary missing value (.) so that it can distinguish calculated results that happen to be missing from results that have not yet been calculated.

Here is the entire code for calculating x, y, and z from a and b:

```
struct info
{
      // ----------------------- input variables ---
      real scalar   a
      real scalar   b

      //     ==----------------- derived variables ---
      real scalar   x
      real scalar   y
      real scalar   z
}
struct info scalar cset(real scalar a, real scalar b)
{
      struct info scalar   r

      // ------------------- store input variables ---
      r.a =   a
      r.b =   b

      // ----------------- set derived vars to .m ---
      //                   (.m means not yet calculated)
      r.x = r.y = r.z = .m

      return(r)
}
real scalar c_x(struct info scalar r)
{
      if (r.x == .m) {
          r.x = (r.a + r.b) * c_z(r)
      }
      return(r.x)
}
real scalar c_y(struct info scalar r)
{
      if (r.y == .m) {
          r.y = 2 * r.b *c_x(r)
      }
      return(r.y)
}
real scalar c_z(struct info scalar r)
{
      if (r.z == .m) {
          r.z = r.a - r.b
      }
      return(r.z)
}
```

Users begin by calling **cset()** to specify the values of the inputs:

```
: r = cset(3, 4)
```

`cset()` also sets the derived variables to `.m` to mark them as not yet calculated.

Users can then call the `c_*()` functions. They can call them in any order, omit some, and even call others repeatedly:

```
: c_x(r)
 -7
: c_z(r)
 -1
: c_x(r)
 -7
```

The `c_*()` routines are exceedingly smart. They never make an unnecessary calculation, and they never make any calculation twice. They never waste time, and users do not have to tell them what is wanted ahead of time or run the routines in a specified order.

We will use this approach for writing linear regression. Some users may want just the coefficients; others, the coefficients and variance matrix; and still others, everything. All will be satisfied. Users will even be able to specify a regression model and obtain just the R-squared if that is their desire. And the R-squared will be calculated as efficiently as possible.

There are variations on how self-threading code is written for special cases. Related values can be grouped and calculated together if that is convenient. If derived variables u, v, and w are nearly always used together, or if they are easier to calculate jointly, you can write code that calculates all three when any one is requested:

```
real whatever c_u(struct info scalar r)
{
        if (r.u == .m) calc_uvw(r)
        return(r.u)
}
real whatever c_v(struct info scalar r)
{
        if (r.v == .m) calc_uvw(r)
        return(r.v)
}
real whatever c_w(struct info scalar r)
{
        if (r.w == .m) calc_uvw(r)
        return(r.w)
}
void calc_uvw(struct info scalar r)
{
        r.u = ...
        r.v = ...
        r.w = ...
}
```

Another variation deals with lightweight calculations. If some formulas are quick to calculate, you can write their code to make their calculation repeatedly rather than storing the result:

```
real whatever c_simple(struct info scalar r)
{
        return(calculation)
}
```

You might even use this style when a calculation is time consuming but rarely called, especially if storing the result consumes lots of memory.

A third variation deals with different formulas for making the same generic calculation. Stata's **regress** command, for instance, provides traditional variance estimates and robust variance estimates. Other formulas depend on the variance estimates. Some can use either, while others must use one or the other. In this case, you create three calculation functions:

Function c_V() returns whichever variance estimate the caller has previously set as the default.

Function c_V_td() returns the traditional estimates.

Function c_V_robust() returns the robust estimates.

Functions of other statistics call c_V(), c_V_td(), or c_V_robust() as appropriate.

The code for the variance functions is

```
real matrix c_V(struct info scalar r)
{
        return( r.Vmethod==1 ? c_V_td(r)  : c_V_robust(r) )
}
real matrix c_V_td(struct info scalar r)
{
        if (r.V_td == .m) {
            r.V_td = ...
        }
        return(r.V_td)
}
real matrix c_V_robust(struct info scalar r)
{
        if (r.V_robust == .m) {
            r.V_robust = ...
        }
        return(r.V_robust)
}
```

Putting the above aside, there is only one more thing you need to know, and that concerns the storing of input variables. I oversimplified the issue. The structure definition I showed was

```
struct info
{
        // ------------------------- input variables ---
        real scalar    a       //   <-- problem here
        real scalar    b       //   <-- problem here
        // --------------------- derived variables ---
        real scalar    x
        real scalar    y
        real scalar    z
}
```

What I showed was fine for dealing with scalars. Had the inputs been vectors, matrices, or structures, however, it would have been better to save pointers to them instead of copies, as was explained in section 10.10. Doing that saves memory. If inputs a and b were matrices instead of scalars, it would be better if the code read

```
struct info
{
        // ------------------------- input variables ---
        pointer(real matrix) scalar    a
        pointer(real matrix) scalar    b
        // --------------------- derived variables ---
        real matrix    x
        real matrix    y
        real matrix    z
}
struct info scalar cset(real matrix a, real matrix b)
{
struct info scalar    r
        // ------------------- store input variables ---
        r.a =    &a
        r.b =    &b
        // ----------------- set derived vars to .m ---
        //                   (.m means not yet calculated)
        r.x = r.y = r.z = .m
}
```

The original version of the c_z() function read

```
real scalar c_z(struct info scalar r)
{
        if (r.z == .m) {
            r.z = r.a - r.b
        }
        return(r.z)
}
```

Because **r.a** and **r.b** are now pointers, the updated version would read

```
real matrix c_z(struct info scalar r)
{
        if (r.z == .m) {
            r.z = *r.a - *r.b
        }
        return(r.z)
}
```

11.3 Linear-regression system lr*() version 1

We are about to write a linear-regression system using self-threading code, which means that we will have lots of subroutines. To prevent them from being confused with subroutines of other systems, I decided that we would use subroutine names starting with the letters **lr**. They would have names such as **lr_b()**, **lr_V()**, and so on. As I wrote the code, I found myself thinking of the routines as the **lr_*()** routines. I decided to call the system the **lr*()** functions, pronounced *"el are star"*.

The **lr*()** functions will fit models of the form

$$\mathbf{y} = \mathbf{X}\boldsymbol{\beta} + \mathbf{e} \qquad\qquad \text{(models without intercepts)}$$

$$\mathbf{y} = \mathbf{X}\boldsymbol{\beta} + c + \mathbf{e} \qquad\qquad \text{(models with intercepts)}$$

Users of **lr*()** will first call **lrset()** to specify **y**, **X**, and whether to include an intercept. They will then call other **lr*()** functions to obtain whichever results they desire.

Obviously, I have already written **lr*()**, but before I show you the code, let me show you **lr*()** in action. I think you will be impressed, especially when I tell you that the entire system consists of only 125 lines of code, excluding comments and blank lines.

11.3.1 lr*() in action

Users of **lr*()** call **lrset()** to specify their regression problem:

```
r = lrset(y, X, 1)
```

lrset() takes three arguments: a column vector containing the values of the dependent variable, a matrix containing the values of the independent variables, and 1 or 0 specifying whether the model is to include an intercept.

lrset() returns **r**, which happens to be a structure. I say happens to be because I do not intend that users access the structure's members. This makes **r** a passthru variable (see section 6.3.2.1), and I might not even tell **lr*()** users that **r** is a structure. I could document that **r** is transmorphic.

Users start by specifying the regression problem with **r** = **lrset()**, and after that, they can obtain coefficients by coding

```
b = lr_b(r)
```

or variance estimates by coding

```
V = lr_V(r)
```

or fitted values by coding

```
yhat = lr_yhat(r)
```

and so on. Let me show you. I have already defined a **y** vector and **X** matrix in Mata:

```
: y
```

	1
1	22
2	17
3	22
4	20
5	15
6	18
7	26
8	20
9	16
10	19

```
: X
```

	1	2
1	2930	121
2	3350	258
3	2640	121
4	3250	196
5	4080	350
6	3670	231
7	2230	304
8	3280	196
9	3880	231
10	3400	231

To fit the model $\mathbf{y} = \boldsymbol{\beta} + c + \mathbf{e}$ and obtain **b**, the fitted values of $(\boldsymbol{\beta} \setminus c)$, and the variance matrix of **V**, I will type

```
: r = lrset(y, X, 1)
: b = lr_b(r)
: V = lr_V(r)
```

I do that below and display the contents of **b** and **V**:

```
: r = lrset(y, X, 1)        // set problem
: b = lr_b(r)               // obtain coefficient vector
: b
                   1

    1      -.0056644997
    2       .0004631627
    3       37.92487622


: V = lr_V(r)               // obtain variance matrix
: V
[symmetric]
                   1                  2                  3

    1       4.17119e-07
    2      -1.16730e-06        .0000252448
    3      -.0011030394       -.0018340879       4.120595787
```

lr*() is certainly easy to use.

In what I ran, however, I violated the spirit of self-threading code by storing **lr_b()**'s and **lr_V()**'s returned values in **b** and **V**. Storing them was wasteful of memory and unnecessary because self-threading systems are designed so that their functions can be called repeatedly with no performance penalty. If I wanted to show the values, I should have just coded

```
: r = lrset(y, X, 1)
: lr_b(r)
: lr_V(r)
```

In my defense, the `lr*()` functions are not intended for interactive use. Mata itself is not intended for interactive use, or at least not intended for interactive uses except for programmers to experiment and authors to present demonstrations. Let's look at how a programmer might use `lr_b()` and `lr_V()`. The program below obtains and displays regression results:

```
: void displayresults(transmorphic r)
> {
>         real scalar    i
>         string scalar name
>
>         for (i=1; i<=rows(lr_b(r)); i++) {
>              name = (i==rows(lr_b(r)) ? "intercept" :
>                                         "x" + strofreal(i))
>                 printf("   %10s  %9.0g %9.0g\n",
>                                 name,
>                                 lr_b(r)[i],
>                                 sqrt(lr_V(r)[i,i])
>                            )
>         }
> }

: displayresults(r)
             x1  -.0056645   .0006458
             x2   .0004632   .0050244
       intercept   37.92488   2.029925
```

`displayresults()` repeatedly calls `lr_b()` and `lr_V()`. One place the functions appear is in the `printf()` statement inside the `for` loop. The statement is

```
printf("   %10s  %9.0g %9.0g\n",
                name,
                lr_b(r)[i],
                sqrt(lr_V(r)[i,i])
           )
```

The code prints lines such as

```
   x1  -.0056645   .0006458
```

The numeric values shown correspond to `lr_b(r)[i]` and `sqrt(lr_V(r)[i,i])`.

The `[i]` and `[i,i]` are subscripts. They are the same subscripts that might be used to subscript variables such as `b[i]` or `V[i,i]`. Results returned by functions can be subscripted just as variables can be subscripted. `lr_b(r)` returns a vector, and thus `lr_b(r)[i]` is the ith element of the `lr_b(r)` vector.

This usage might bother you because it seems inefficient. Are we really going to fetch the entire `lr_b(r)` vector just to obtain one element from it, and then next time through the loop, do it all over again? Yes, that is precisely what we are going to do, and no, it is not inefficient. It is not even inefficient that the `for` loop itself obtains `rows(lr_b(r))` repeatedly merely to determine whether the loop should continue:

```
for (i=1; i<=rows(lr_b(r)); i++) {
        ...
        printf("   %10s   %9.0g %9.0g\n",
                        name,
                        lr_b(r)[i],
                        sqrt(lr_V(r)[i,i])
                )
}
```

Function calls consume barely more time than variable references if we ignore the time spent executing the function. Self-threading functions execute in nearly zero time when they have previously calculated the result. The above is how you are supposed to use self-threading functions.

I have not shown you all of `lr*()`'s capabilities. I included a handful of other functions to calculate other results, and still more need to be added. Available are

```
: lr_ee(r)        // error sum of squares
  7.133114624

: lr_s2(r)        // variance of e
  1.019016375

: lr_N(r)         // # of observations
  10

: lr_k(r)         // # of RHS vars, excluding intercept
  2

: lr_K(r)         // # of RHS vars, including intercept
  3

: lr_K_adj(r)     // lr_K() adjusted for dropped (collinear) vars
  3

: lr_yhat(r)      // predicted values
                     1
```

```
      1  |   21.38393493
      2  |   19.06829838
      3  |   23.02663983
      4  |   19.60603225
      5  |    14.9758246
      6  |   17.24315309
      7  |   25.43384347
      8  |   19.43609726
      9  |   16.05360817
     10  |    18.772568
```

The `lr*()` functions can be called in any order.

Consider the programmer who needs only the fitted values. That programmer could code

```
r = lrset(y, X, 1)
lr_yhat(r)
```

`lr*()` is truly programmer friendly.

11.3.2   The calculations to be programmed

Before you write code, you must know what the code needs to do. In statistical applications, this means identifying the formulas.

I do not want you to get lost in the formulas required to implement linear regression. This book is first and foremost a programming book, albeit for programmers interested in adding features to Stata. Even so, I will discuss the formulas a little because I know some readers will be interested.

What every reader needs to understand is that the formulas need to be written down before you write a line of code. At StataCorp, we write them on formula sheets which, despite the officious name, are just pieces of paper. They are, however, important pieces of paper.

I wrote a formula sheet for `lr*()`. Some of the formulas I wrote have lousy numerical properties. We will substitute better formulas for them later. It will not be difficult to do, because self-threading code is so easily modified. I did not, however, write lousy formulas just so I could show you how to modify them later. I write lousy formulas and substitute better ones later even when I am not writing code. I like to make systems work first and substitute better formulas later, because better formulas are usually just one more thing to go wrong and debug.

If you really do try to follow the formulas listed below, watch out for k and K. Lowercase k is the total number of independent variables excluding the constant even if there is one. Capital K includes the constant and so equals k or k+1.

Finally, most researchers use the words *constant* and *intercept* interchangeably. I sometimes do that myself. Nonetheless, the intercept is a coefficient, and the constant is the vector of 1s appended to the outside of X so that the intercept can be fit.

The formulas are

── Formula sheet 1 ─────────

Model to be fit:

    (1) y = Xb + e   model without intercept
    (2) y = (X,1)*b + e   model with intercept
                                                (1 is colvector of 1s)

where

    y: N x 1   obs. on dependent variables
    X: N x k   obs. on k independent variables excluding constant

Define

    cons = 0 or 1   meaning fit model 1 or fit model 2
       k = cols(X)   = # of independent variables excluding constant
       K = k + cons = # of independent variables including constant
       N = rows(X)   = # of observations

       K_adj   = K adjusted for variables dropped because of
                           collinearity. The i-th diagonal element
                           of invsym(X'X) is 0 when variable is
                           dropped. Thus, K_adj is K minus
                           the number of 0s along diagonal.

Thus,

    b: k   x 1   if model (1)
        k+1 x 1   if model (2)
         K   x 1   in both cases

The linear-regression formulas are

     b = invsym(XX)*Xy   lousy numerical properties!

    XX = X'X   X augmented with 1s if intercept

    Xy = X'y   X augmented with 1s if intercept

    yy = y'y

    e'e = yy - b'X'Xb   error sum of squares

    V = s2*invsym(XX)   variance matrix (VCE)

    s2 = e'e / (N-K_adj)

                            predicted values
    yhat = X*b   if r.cons==0
        = (X,1)*b   if r.cons==1

── Formula sheet 1 ─────────

11.3.3   lr*() version-1 code listing

Before I talk you through the code, you need to see it for yourself. I want you to appreciate just how readable self-threading code is. The main routines are

```
r = lrset(y, X, cons)
lr_b(r)
lr_V(r)
```

Find those routines and use them as anchors. I doubt you will get lost. There are only 125 lines, omitting comments and blanks.

By the way, I used **.z** instead of **.m** to mark not-yet-calculated results. Either value works equally well.

You can see the full listing of the **lr*()** code by typing in Stata

```
. view ~/matabook/lr1.mata
```

If you have not yet downloaded the files associated with this book, see section 1.4. It is so important that you look at the entire code, however, that I list it below. I told you about the order in which I wrote the routines. Yet the routines listed below are in a different order! Of course they are. I reorganized the files.

── lr1.mata ─────────

```
 1:  *! lr*() version 1.0.0
 2:
 3:  version 15
 4:  set matastrict on
 5:
 6:  mata:
 7:
 8:  struct lrinfo
 9:  {
10:          // -------------------------------------- inputs
11:          pointer(real colvector) scalar    y
12:          pointer(real matrix) scalar       X
13:          real scalar                       cons
14:
15:          // -------------------------------------- derived
16:          /*
17:              Notation:
18:                  y = Xb + e
19:                yhat = Xb
20:
21:                  N = # of obs
22:                  K = # of indep vars including intercept if present
23:                  k = # of indep vars excluding intercept
24:
25:              K_adj = K adjusted for collinearity
26:          */
27:
```

```
28:            real matrix       XX             // X´X; K x K
29:            real colvector    Xy             // X´y; K x 1
30:            real matrix       XXinv          // (X´X)^(-1);   K x K
31:            real colvector    b              // coefficients; K x 1
32:            real matrix       V              // variance matrix (VCE); K x K
33:            real scalar       s2             // variance of e
34:            real scalar       yy             // y´y
35:            real scalar       ee             // e´e, error sum of squares
36:            real scalar       K_adj
37:
38:            // ---------------------- derived but not stored
39:    }
40:
41:    void lrinit(struct lrinfo scalar r)
42:    {
43:            r.XX      = .z
44:            r.Xy      = .z
45:            r.XXinv   = .z
46:            r.b       = .z
47:            r.V       = .z
48:            r.s2      = .z
49:            r.yy      = .z
50:            r.ee      = .z
51:            r.K_adj   = .z
52:    }
53:
54:
55:    /* ----------------------------------------- entry points --- */
56:
57:    struct lrinfo scalar lrset( real colvector y,
58:                               real matrix    X,
59:                               real scalar    cons )
60:    {
61:            struct lrinfo scalar   r
62:
63:            assert(rows(y)==rows(X))
64:
65:            r.y    = &y
66:            r.X    = &X
67:            r.cons = (cons!=0)
68:
69:            lrinit(r)              // initialize derived vars
70:
71:            return(r)
72:    }
73:
74:    real scalar lr_N(struct lrinfo scalar r) return(rows(*r.X))
75:    real scalar lr_k(struct lrinfo scalar r) return(cols(*r.X))
76:    real scalar lr_K(struct lrinfo scalar r) return(cols(*r.X) + r.cons)
77:
78:
79:    real scalar lr_K_adj(struct lrinfo scalar r)
80:    {
81:            if (r.K_adj == .z) {
82:                    r.K_adj = colsum(diagonal(lr_XXinv(r)) :!= 0)
83:            }
84:            return(r.K_adj)
85:    }
```

```
 86:
 87:
 88:  real colvector lr_b(struct lrinfo scalar r)
 89:  {
 90:          if (r.b == .z) {
 91:                  r.b = lr_XXinv(r)  * lr_Xy(r)
 92:          }
 93:          return(r.b)
 94:  }
 95:
 96:  real matrix lr_V(struct lrinfo scalar r)
 97:  {
 98:          if (r.V == .z) {
 99:                  r.V = lr_s2(r)  * lr_XXinv(r)
100:          }
101:          return(r.V)
102:  }
103:
104:  real scalar lr_s2(struct lrinfo scalar r)
105:  {
106:          if (r.s2 == .z) {
107:                  r.s2 = lr_ee(r) / ( lr_N(r) - lr_K_adj(r) )
108:          }
109:          return(r.s2)
110:  }
111:
112:  real scalar lr_ee(struct lrinfo scalar r)
113:  {
114:          if (r.ee == .z) {
115:                  r.ee = lr_yy(r) - lr_b(r)'lr_XX(r)*lr_b(r)
116:          }
117:          return(r.ee)
118:  }
119:
120:
121:  real colvector lr_yhat(struct lrinfo scalar r)
122:  {
123:          return( (*r.X)*lr_b_X(r) :+ lr_b_c(r) )
124:  }
125:
126:  real colvector lr_b_X(struct lrinfo scalar r)
127:  {
128:          if (r.cons) {
129:                  if (lr_k(r)) return( lr_b(r)[| 1 \ lr_k(r) |])
130:                  else         return( J(0, 1, .))
131:          }
132:          else               return(lr_b(r))
133:  }
134:
135:  real scalar lr_b_c(struct lrinfo scalar r)
136:  {
137:          return( r.cons ? lr_b(r)[lr_K(r)] : 0 )
138:  }
139:
140:
```

```
141:    /* -------------------------------internal subroutines --- */
142:
143:    real matrix lr_XX(struct lrinfo scalar r)
144:    {
145:            if (r.XX == .z) {
146:                    r.XX = cross(*r.X, r.cons,  *r.X, r.cons)
147:            }
148:            return(r.XX)
149:    }
150:
151:    real colvector lr_Xy(struct lrinfo scalar r)
152:    {
153:            if (r.Xy == .z) {
154:                    r.Xy = cross(*r.X, r.cons,  *r.y, 0)
155:            }
156:            return(r.Xy)
157:    }
158:
159:    real scalar lr_yy(struct lrinfo scalar r)
160:    {
161:            if (r.yy == .z) {
162:                    r.yy = cross(*r.y, *r.y)
163:            }
164:            return(r.yy)
165:    }
166:
167:
168:    real matrix lr_XXinv(struct lrinfo scalar r)
169:    {
170:            if (r.XXinv == .z) {
171:                    r.XXinv = invsym(lr_XX(r))
172:            }
173:            return(r.XXinv)
174:    }
175:
176:    end
```

── lr1.mata ────────

11.3.4   Discussion of the lr*() version-1 code

Even if you looked at the code carefully, I bet it escaped your attention that it handles models with no independent variables and models fit on zero observations. I am nearly certain you did not notice because there was nothing to notice. Mata itself handled the problem because it allows null matrices. If the model has no independent variables, then the matrix $\mathbf{X}$ (*r.X in the code) is $N \times 0$. If there are no observations, then *r.X is $0 \times k$.

The code does not handle missing values, however. If *r.y or *r.X contain missing values, the code does not crash, but it does produce incorrect results. This is acceptable as long as callers are warned, just as it was acceptable for Mata's invsym() function to produce incorrect results with nonsymmetric matrices. Both routines are intended for use by programmers. The user of lr*() can verify that (*r.y) and (*r.X) do not contain missing values easily enough:

```
if (missing(y) + missing(X)) ...
```

Even so, `lr*()` would have handled missing values if I had written it for my day job. I did not handle them because it would have made the code more confusing, and I did not want you to become immersed in a detail. Here is how you could fix the problem. Although I said the code does not handle missing values, it mostly handles them. Most routines use `cross()` for forming matrix cross products, and `cross()` handles missing values automatically. `cross()` was not used, however, in coding `lr_ee()` (lines 112–118 of `lr1.mata`). It needs to be rewritten. The only other problem is in `lr_N()` (line 74). Its code needs to be modified to return the number of complete observations.

Let me talk you through parts of the code.

11.3.4.1 Getting started

The set-up routine is named `lrset()` and reads

```
struct lrinfo scalar lrset( real colvector y,
                            real matrix    X,
                            real scalar    cons )
{
        struct lrinfo scalar    r
        assert(rows(y)==rows(X))
        r.y    = &y
        r.X    = &X
        r.cons = (cons!=0)
        lrinit(r)                  // initialize derived vars
        return(r)
}
```

`r.cons` is set equal to `cons!=0` instead of `cons` to ensure that `r.cons` will be 0 or 1. Remember that Mata allows logical variables to be zero or not zero. The user might have specified `cons` equal to three. I intend to use `r.cons` not only in if statements but also in formulas such as `K = k + r.cons` where it is important that `r.cons` be 0 or 1.

I next wrote three short functions:

```
real scalar lr_N(struct lrinfo scalar r) return(rows(*r.X))
real scalar lr_k(struct lrinfo scalar r) return(cols(*r.X))
real scalar lr_K(struct lrinfo scalar r) return(cols(*r.X)
                                                    + r.cons)
```

Did you know that you can omit the braces around the body of functions that consist of just one statement?

11.3.4.2 Assume subroutines

The first substantive program I wrote was `lr_b()`, the routine for obtaining the regression coefficients. It is

```
real colvector lr_b(struct lrinfo scalar r)
{
        if (r.b == .z) {
            r.b = lr_XXinv(r) * lr_Xy(r)
        }
        return(r.b)
}
```

The line

```
r.b = lr_XXinv(r) * lr_Xy(r)
```

is a direct translation of the formula $\mathbf{b} = (\mathbf{X'X})^{-1}\mathbf{X'y}$. I decided that `lr_XXinv()` would return $(\mathbf{X'X})^{-1}$ and that `lr_Xy()` would return $\mathbf{X'y}$.

I write systems from the top down. As I write code, I assume the existence of subroutines that return what I need in the form that I need it. I assume solutions even though I have yet to write them. When I later write the subroutines, I will do the same again if that is convenient, resulting in yet more subroutines to write later. In this way, details are pushed down until it is convenient for me to deal with them.

I want you to appreciate how many details are being hidden by the line

```
r.b = lr_XXinv(r) * lr_Xy(r)
```

How will `lr_XXinv()` obtain $(\mathbf{X'X})^{-1}$? Presumably by taking the `invsym()` of $\mathbf{X'X}$. `invsym()` is capable of handling collinearity when the matrix is not full rank. How will $\mathbf{X'X}$ be calculated? I do not know. What about models with and without an intercept? I do not know. How will the system obtain $\mathbf{X'X}$ for `invsym()` to invert when $\mathbf{X}$ does not contain a column of 1s and needs to contain them? I do not know.

Well, I do know, but it would not bother me if I did not. I will handle the problems when I come to them.

In fact, I postponed the details of how to calculate $(\mathbf{X'X})^{-1}$ yet again when I wrote the routine to make the variance calculation, `lr_V()`. It is

```
real matrix lr_V(struct lrinfo scalar r)
{
        if (r.V == .z) {
            r.V = lr_s2(r) * lr_XXinv(r)
        }
        return(r.V)
}
```

The code for `lr_V()`, like the code for `lr_b()`, is right from the formula sheet and, just as I did when writing `lr_b()`, I assumed subroutines for its ingredients. `lr_s2()` is yet another function I will have to write later. Meantime, the assuming of subroutines is already beginning to pay off. This is the second time we have used `lr_XXinv()`.

I may as well write `lr_XXinv()` now.

```
real matrix lr_XXinv(struct lrinfo scalar r)
{
        if (r.XXinv == .z) {
            r.XXinv = invsym(lr_XX(r))
        }
        return(r.XXinv)
}
```

Mata's built-in `invsym()` function handles collinearity, but notice I still have not dealt with how $\mathbf{X'X}$ will be obtained when there is a constant. I postponed that problem by coding `lr_XX(r)`, but I have made an assumption. `lr_XX()` will return $(\mathbf{X}, \mathbf{1})'(\mathbf{X}, \mathbf{1})$ when the model includes an intercept.

I next coded `lr_s2()`:

```
real scalar lr_s2(struct lrinfo scalar r)
{
        if (r.s2 == .z) {
            r.s2 = lr_ee(r) / ( lr_N(r) - lr_K_adj(r) )
        }
        return(r.s2)
}
```

Are you keeping track of what I have yet to write? I am not. I just look back at what I have written and pick an unwritten function. If I forget to write a function, Mata will remind me when I try to compile and execute the code.

11.3.4.3   Learn about Mata's built-in subroutines

I decided to write `lr_K_adj()` next. `lr_K_adj()` is an interesting function. It is the number of independent variables excluding variables dropped because of collinearity. Variables were dropped because of collinearity by `lr_Xinv()` when it used `invsym()` to invert the matrix. Somehow, we are going to have to determine how many variables it dropped, if any.

We have three k's floating around in the code: `lr_k()`, `lr_K()`, and `lr_K_adj()`.

`lr_k()` is the total number of independent variables. It was one of the one-line functions that I wrote earlier (line 75):

```
real scalar lr_k(struct lrinfo scalar r) return(cols(*r.X))
```

lr_K() is the number of independent variables including the constant if there is one. It
was another of the one-line functions (line 76):

```
real scalar lr_K(struct lrinfo scalar r) return(cols(*r.X)
                                                    + r.cons)
```

lr_K_adj() will be lr_K() minus the number of variables dropped because of collinearity.
I need to count the number of dropped variables.

As I said, the variables were dropped by **lr_XXinv()** when it used **invsym()** to invert
the matrix. The way **invsym()** works, if variable i was dropped, then **lr_XXinv()**$[i,i]$
was set to the specified 0. The total number of dropped variables is thus the number
of 0s along the diagonal of **lr_XXinv()**. Equivalently, the number of included variables
is the number of nonzero values along the diagonal. Here is code to calculate that:

```
real scalar lr_K_adj(struct lrinfo scalar r)
{
        if (r.K_adj == .z) {
            r.K_adj = colsum(diagonal(lr_XXinv(r)) :!= 0)
        }
        return(r.K_adj)
}
```

The key part of the calculation is

```
r.K_adj = colsum(diagonal(lr_XXinv(r)) :!= 0)
```

Mata's **diagonal()** function extracts the diagonal of a matrix into a column vector.
The colon-operator **:!=** in

```
diagonal(lr_XXinv(r)) :!= 0
```

compares each element of the vector with 0 and returns a column vector of 0s and 1s,
meaning not equal or equal to 0. Mata's **colsum()** function sums that vector, directly
yielding the number of nonzero elements.

I am obviously adept in the use of Mata's built-in functions. I do not expect you to be as
adept with them yet. You do, however, need to become familiar with them. I recommend
you type **help mata**, click on **[M-4] Categorical guide to Mata functions**, and begin
exploring the online documentation. Otherwise, you will end up writing code that looks
like this:

```
real scalar lr_K_adj(struct lrinfo scalar r)
{
        real scalar    i
        if (r.K_adj == .z) {
            r.K_adj = 0
            for (i=1; i<=rows(lr_XXinv(r)); i++) {
                    if (lr_XXinv(r)[i,i]) {
                            r.K_adj = r.K_adj + 1
                    }
            }
        }
        return(r.K_adj)
}
```

It would not be horrible if you wrote code like the above. It would execute more slowly than my code, but I doubt anyone would notice, and the code is certainly clear about what it does. Nevertheless, it was easier for me to write

```
r.K_adj = colsum(diagonal(lr_XXinv(r)) :!= 0)
```

There are certain Mata functions that I want to emphasize to you. `invsym()` is one because you obviously do not want to write your own matrix inverter.

`cross()` is another, which we will discuss in the next section.

11.3.4.4 Use of built-in subroutine cross()

Calculations such as X'X, X'W*X, and X'diag(w)*X often arise in statistical calculations. You can program them just as I have written them, or you can use Mata's `cross()` function. Using `cross()` has advantages. `cross()` uses less memory, it is sometimes faster, and it handles missing values.

I used `cross()` to program the calculations of $\mathbf{X'X}$, $\mathbf{X'y}$, and $\mathbf{y'y}$ in lr_XX(), lr_Xy(), and lr_yy(). If you look back at the code listing (lines 146, 154, and 162), the calls those functions make to `cross()` are

```
r.XX = cross(*r.X, r.cons,   *r.X, r.cons)
r.Xy = cross(*r.X, r.cons,   *r.y,      0)
r.yy = cross(*r.y,           *r.y        )
```

cross(A, B) calculates $A'B$. Thus,

```
r.yy = cross(*r.y,           *r.y        )
```

calculates (*r.y)'(*r.y). `cross()` uses less memory to make the calculation.

I could have used cross(*r.X, *r.X) to calculate X'X. If you look back to the formula sheet, however, you will see the line,

```
XX = X'X                        X augmented with 1s if intercept
```

If you do not know what "augmented with 1s" means, I apologize. You write formula sheets so that you—the programmer writing the code—can understand them, and I knew exactly what I meant by "augmented with 1s". What the formula sheet was saying is that the **XX** is

```
if (r cons) {
        One = J(1, rows(X), 1)    // N x 1 vector of 1s
        XX  = (X,One)´(X,One)
}
else    XX  = X´X
```

Wanting to augment a matrix with 1s occurs so often that **cross()** has a special syntax for it. **cross()** has two syntaxes, one with two arguments and another with four arguments:

cross(X , X)	calculates	X'X
cross(X,0, X,0)	calculates	X'X
cross(X,1, X,0)	calculates	(X,One)'X
cross(X,0, X,1)	calculates	X'(X,One)
cross(X,1, X,1)	calculates	(X,One)'(X,One)

Thus, to calculate **r.XX**, I used

```
r.XX = cross(*r.X, r.cons,  *r.X, r.cons)
```

To calculate **r.Xy**, I used

```
r.Xy = cross(*r.X, r.cons,  *r.y,      0)
```

Here are the routines:

```
real matrix lr_XX(struct lrinfo scalar r)
{
        if (r.XX == .z) {
            r.XX = cross(*r.X, r.cons,  *r.X, r.cons)
        }
        return(r.XX)
}

real colvector lr_Xy(struct lrinfo scalar r)
{
        if (r.Xy == .z) {
            r.Xy = cross(*r.X, r.cons,  *r.y, 0)
        }
        return(r.Xy)
}
```

```
real scalar lr_yy(struct lrinfo scalar r)
{
        if (r.yy == .z) {
            r.yy = cross(*r.y, *r.y)
        }
        return(r.yy)
}
```

You will be seeing `cross()` again later in this chapter. `cross()` comes in four flavors:

`cross()`	double precision
`quadcross()`	quad precision
`crossdev()`	double precision, deviation form
`quadcrossdev()`	quad precision, deviation form

11.3.4.5   Use more subroutines

`lr_yhat()` is worthy of comment because of the subroutines I wrote to make the calculation easier to program. Predictions are easy to calculate. Quoting from the formula sheet, they are

```
yhat = X*b              if r.cons==0
     = (X,1)*b          if r.cons==1
```

First, I guess I should apologize again. `(X,1)` is a shorthand way to write "augmented with 1s". It is a shorthand that humans understand, but Mata does not. Second, note that `b` has an extra element—the intercept—if `r.cons==1`. I decided to create subroutines so that, whether `r.cons` was 0 or 1, I could write the predicted values as

```
yhat = X*lr_b_X() :+ lr_b_c()
```

`lr_b_X()` would be `X*b` ignoring the intercept in all cases, and `lr_b_c()` would be the intercept or 0. Function `lr_yhat()` thus became

```
real colvector lr_yhat(struct lrinfo scalar r)
{
        return( (*r.X)*lr_b_X(r) :+ lr_b_c(r) )
}
```

`lr_b_X()` would return the first `lr_k()` elements of `lr_b()`. `lr_b_c()` would return either `lr_b()[lr_K()]` or 0.

Function `lr_b_c()` was easy enough to code:

```
real scalar lr_b_c(struct lrinfo scalar r)
{
        return( r.cons ? lr_b(r)[lr_K(r)] : 0 )
}
```

For `lr_b_X()`, I coded

```
real colvector lr_b_X(struct lrinfo scalar r)
{
        if (r.cons) {
                if (lr_k(r)) return( lr_b(r)[| 1\lr_k(r) |])
                else         return( J(0, 1, .))
        }
        else                 return(lr_b(r))
}
```

11.4 Linear-regression system lr*() version 2

I warned you that I used formulas in `lr*()` version 1 that have lousy numerics. We are going to fix that. Before we do, however, let me explain how I write code. I use second-rate formulas or methods in my first drafts because that makes the initial programming task easier. I splice in better routines once the system is working.

Below is a table summarizing where we are and where we are going. There is a long note at the foot of the table. The note is longer than it is important. The table reports digits of accuracy, and the note states how that accuracy is measured. The table is

Method		Base-10 digits of accuracy	
Absolute	`cross()`	5.8	`lr*()` version 1
Absolute	`quadcross()`	7.2	
Deviation	`cross()`	11.8	
Deviation	`quadcross()`	13.1	`lr*()` version 2

Reported digits of accuracy are the average digits of accuracy over 10,000 regressions. For each regression, accuracy is measured as $-\text{log10}(\textit{relative difference})$ between the most inaccurately calculated coefficient and the coefficient that would have been obtained had the calculation been performed with infinite precision.

The 10,000 regressions were run on 10,000 different datasets. Each dataset contained 2,000 observations on y, x_1, x_2, and x_3, where $y = 1 * x_1 + 2 * x_2 + 3 * x_3 + 52 + e$, with e a normal random variate with mean 0 and standard deviation 8. Variables x_1, x_2, and x_3 were correlated with each other 0.9 (expected) and had means 50, 2,000, and $-1,000$ (expected). The average r^2 of the 10,000 regressions was 0.35.

Where we are is 5.8 digits of accuracy. Where we are going is 13.1, a highly respectable neighborhood. The theoretical maximum accuracy is 15.7 digits, and that is unachievable.

We will make two changes to `lr*()` to get from 5.8 to 13.1. We will change every call of `cross()` to `quadcross()`, the version of `cross()` that performs its calculation in quad precision. That by itself will take us to 7.2 digits of accuracy.

We will also substitute better formulas for fitting linear regressions. That will be more work, but it will take us to 11.8 digits of accuracy all by itself.

The two changes together will take us all the way to 13.1 digits.

11.4.1 The deviation from mean formulas

The inaccuracy of results in the current code arises because of numerical problems in calculating $\mathbf{b} = (\mathbf{X}'\mathbf{X})^{-1} * \mathbf{X}'\mathbf{y}$. The problems are not so much with the formula itself as they are with its ingredients, $\mathbf{X}'\mathbf{X}$ and $\mathbf{X}'\mathbf{y}$. Calculating them requires performing lots of multiplications and additions.

Computers perform multiplication and division accurately—you can depend on 15.7 digits of accuracy. The same is not true of addition and subtraction. Calculated results can vary from 15.7 to 0 digits of accuracy. As a general rule, the farther apart the values being added or subtracted, the worse will be the accuracy.

One job of numerical analysts is to find transforms that move values to be added and subtracted closer to each other. Not any transform will do. The idea is to transform the input values, make the calculations on the transformed values (sometimes the formulas are modified, too), and finally reverse-transform the calculated results. Transforms that can be undone at the last step do not always exist.

Transforms exist for linear regression, but only for models that include an intercept. $\mathbf{X}$ and $\mathbf{y}$ can be rescaled to have mean 0 which will make values closer to each other. Working with deviations from the means requires modifying a formula here and there, but there is not much that changes.

Here is the updated formula sheet:

———————————————————————————————— Formula sheet 2 ———————

Assume

```
X:   N x k   data matrix
y:   N x 1   dependent variable
k = cols(X)  = # of independent vars excluding intercept
K = k + 1    = # of independent vars including intercept
N = rows(X)  = # of observations
K_adj = colsum(diagonal(SSinv)!:=0)              (1 x 1)
        (K adjusted for dropped variables)
```

Then

```
mean(X) = column means of X                       (1 x k)
mean(y) = column mean  of y                        (1 x 1)
     SS = (X:-mean(X))´(X:-mean(X))               (k x k)
     St = (X:-mean(X))*(y:-mean(y)) \ 0           (K x 1)
  SSinv = ( ((SS)^-1  v), (v´\d) )                (K x K)
      v = -mean(X)*(SS)^-1                         (1 x k)
      d = 1/N - mean(X)*v´                         (1 x 1)
      b = SSinv * st,                              (K x 1)
          replace b[K] = mean(y) - mean(X)*b_X
    b_X = b[| 1\k |]                               (k x 1)
    b_c = b[K]                                     (1 x 1)
    e´e = t´t - b_X´SS*b_X                         (1 x 1)
     s2 = e´e / (N - K_adj)                        (1 x 1)
      V = s2 * SSinv                               (K x K)
   yhat = X*b_X + b_c                              (N x 1)
```

——————————————————————————————— Formula sheet 2 ———————

Note: This formula sheet uses the same notation as the one in section 11.3.2.

11.4.2   The lr*() version-2 code

Modifying the code to use the above formulas resulted in six functions being unchanged, six functions being changed, and six new functions being added. Unchanged are

```
lrset()
lr_XX()
lr_Xy()
lr_yy()
lr_XXinv()
lr_yhat()
```

Changed are

`struct lrinfo`	has new members for SS, St, etc.
`lrinit()`	clears the new members
`lr_K_adj()`	uses SSinv or XXinv as appropriate
`lr_b()`	added new code for r.cons case
`lr_V()`	added new code for r.cons case
`lr_ee()`	added new code for r.cons case

And finally, the new routines are

```
lr_SSinv()
calc_SSinv_cons()
lr_SS()
lr_St()
lr_mean_X()
lr_mean_y()
```

The routines that change do not change much. For instance, `lr_K_adj()` was previously

```
real scalar lr_K_adj(struct lrinfo scalar r)
{
        if (r.K_adj == .z) {
            r.K_adj = colsum(diagonal(lr_XXinv(r)) :!= 0)
        }
        return(r.K_adj)
}
```

and now is

```
real scalar lr_K_adj(struct lrinfo scalar r)
{
        if (r.K_adj == .z) {
            if (r.cons) r.K_adj =
                              colsum(diagonal(lr_SSinv(r)):!=0)
            else            r.K_adj =
                              colsum(diagonal(lr_XXinv(r)):!=0)
        }
        return(r.K_adj)
}
```

In the version-2 code, we must use **lr_SSinv()** in place of **lr_XXinv()** for models with intercepts.

lr_b() changed the most because it must use new formulas for models with intercepts.
lr_b() was previously

```
real colvector lr_b(struct lrinfo scalar r)
{
        if (r.b == .z) {
            r.b = lr_XXinv(r) * lr_Xy(r)
        }
        return(r.b)
}
```

It now is this:

```
real colvector lr_b(struct lrinfo scalar r)
{
        if (r.b == .z) {
            if (r.cons) {
                    r.b            = lr_SSinv(r) * lr_St(r)
                    r.b[lr_K(r)] = lr_mean_y(r) -
                                    lr_mean_X(r)*lr_b_X(r)
            }
            else     r.b = lr_XXinv(r) * lr_Xy(r)
        }
        return(r.b)
}
```

The new routines, meanwhile, calculate the new ingredients that we need. Quoting from formula sheet 2, we need

```
SS = (X:-mean(X))´(X:-mean(X))                    (k x k)
St = (X:-mean(X))*(y:-mean(y)) \ 0               (K x 1)
```

Those formulas, just as they are stated, could be substituted into code. But we are not going to do that. We are going to rewrite them using **cross()** instead. We used **cross()** in the version-1 code for three reasons: 1) it uses less memory, 2) it is sometimes faster, and 3) it handles missing values. There was a fourth reason we used **cross()** that I did not tell you about. `cross()` comes in three other flavors. `quadcross()` is equivalent to `cross()` but performs its calculation in quad precision. `crossdev()` is equivalent to `cross()` but performs its calculation in deviation form. `quadcrossdev()` is equivalent to `crossdev()` but performs its calculation in quad precision.

When I wrote the version-1 code, I knew I was using second-rate formulas, and I knew that I would want to substitute deviation-form calculations. I knew I would want to substitute `crossdev()`. And I knew I would want to perform that calculation in quad precision. I knew I would want to use `quadcrossdev()`.

The calculations we need to code,

```
SS = (X:-mean(X))'(X:-mean(X))
St = (X:-mean(X))*(y:-mean(y)) \ 0
```

can instead be coded

```
SS = quadcrossdev(X,0,mean(X), X,0,mean(X))
St = quadcrossdev(X,0,mean(X), X,0,mean(y)) \ 0
```

Here is why. First, remember that the `cross()` equivalent to A'B is `cross(A, 0, B, 0)`. Therefore, the `cross()` equivalent of

```
SS = (X:-mean(X))'(X:-mean(X))
St = (X:-mean(X))*(y:-mean(y)) \ 0
```

is

```
SS = cross(X:-mean(X),0, X:-mean(X),0)
St = cross(X:-mean(X),0, y:-mean(y),0) \ 0
```

The `crossdev()` equivalent is like `cross()` except it has two more arguments that specify the values to be subtracted:

```
SS = crossdev(X,0,mean(X), X,0,mean(X))
St = crossdev(X,0,mean(X), X,0,mean(y)) \ 0
```

The above is superior to the former because `crossdev()` uses less memory. Finally, to perform the entire calculation in quad precision, we substitute `quadcrossdev()` for `crossdev()`:

```
SS = quadcrossdev(X,0,mean(X), X,0,mean(X))
St = quadcrossdev(X,0,mean(X), X,0,mean(y)) \ 0
```

If you are getting the hang of this, you might wonder why we do not substitute `quadmean()` for `mean()`. We do not because there is no `quadmean()` function. Mata's `mean()` function already works in quad precision. See `help mata mean()`.

The new `lr_SS()` and `lr_St()` routines are

```
real matrix lr_SS(struct lrinfo scalar r)
{
        if (r.SS == .z) {
                r.SS = quadcrossdev(*r.X, 0, lr_mean_X(r),
                                    *r.X, 0, lr_mean_X(r) )
        }
        return(r.SS)
}

real colvector lr_St(struct lrinfo scalar r)
{
        if (r.St == .z) {
                r.St = quadcrossdev(*r.X, 0, lr_mean_X(r),
                                    *r.y, 0, lr_mean_y(r)) \ 0
        }
        return(r.St)
}
```

Rather than using `mean(*r.X)` and `mean(*r.y)` in the calls to `quadcrossdev()`, I used
new functions `lr_mean_X()` and `lr_mean_y()` because we will need the means to implement other formulas, too. The functions follow the self-threading outline of returning
the previously stored value or calculating and storing it:

```
real rowvector lr_mean_X(struct lrinfo scalar r)
{
        if (r.mean_X == .z)  r.mean_X = mean(*r.X)
        return(r.mean_X)
}

real scalar lr_mean_y(struct lrinfo scalar r)
{
        if (r.mean_y == .z)  r.mean_y = mean(*r.y)
        return(r.mean_y)
}
```

11.4.3   lr*() version-2 code listing

You can see the full listing of `lr*()` version-2 code by typing in Stata

```
. view ~/matabook/lr2.mata
```

If you have not yet downloaded the files associated with this book, see section 1.4.

11.4.4   Other improvements you could make

Version 2 is much improved over version 1. At 13.1 digits of accuracy, we are obtaining
83% of the theoretically unobtainable 15.7 digits. I must admit, however, that the
13.1-digit claim is overly optimistic.

Addition and subtraction produce lots of numerical error when the values being added or subtracted are far from each other. Demeaning the variables made the values closer and resulted in improved accuracy.

If we also normalized the variables by their standard deviations, that would make their values even more alike. The accuracy results that I showed you earlier did not reveal that because I set the standard deviations of the independent variables to 1 in the simulations that I ran. Let's find out how much my deception mattered. I reran the simulations, varying the standard deviations. The new results are as follows:

Ratio of largest to smallest SD	Digits of accuracy
1	13.1
1,000	11.3
10,000	10.2
100,000	9.4

SD stands for standard deviation of the independent variables. The first line is the previously reported result.

These results are pretty good. It is a rare real-world regression in which the maximum of the ratios of the standard deviations is more than 10,000, which is why I ran the table out to 100,000. The version-2 code is acceptable even at 100,000 in my view, because 9.4 digits of accuracy is a lot of digits. That said, the standards for judging linear-regression routines are remarkably stringent these days. If you claim to produce professional software and do not get linear regression right, it suggests that you do not get other things right as well.

Normalizing by standard deviations is not difficult. Where the `lr*()` code subtracts the mean, it could divide by each variable's standard deviation, and later in the code, those same standard deviations can be used to undo the adjustment's effect on the calculated coefficients and variance matrix.

There is another improvement we could also make. Rather than obtaining coefficients by calculating $(\mathbf{S'S})^{-1}\mathbf{S't}$, we could solve $(\mathbf{S'S})\mathbf{b} = \mathbf{S't}$ for $\mathbf{b}$. Solvers yield more accurate results. I recommend a solver based on LU decomposition. You could code `b = lusolve(S'S, S't)`; see `help mata lusolve()`. Some experts will tell you to use a more robust solver such as QR decomposition—Mata's `qrsolve()`—but that is not necessary here in my experience.

And of course, we need to add more returned results, meaning more subroutines. Callers will want the F statistic, R-squared, and more. They would be trivial to add.

If any of these projects tempt you, wait. We will re-implement `lr*()` using classes in chapter 13, and the code will become even more elegant and even easier to modify.

11.5   Closeout of lr*() version 2

As it stands, `lr*()` version 2 is excellent code, so we will add it to our `lmatabook.mlib` library.

First, however, the code requires certifying.

11.5.1   Certification

You can see the full listing of `lr*()` version-2 certification script by typing in Stata

```
. view ~/matabook/test_lr2.do
```

Certification is so important that I want to make some comments on what is written in `test_lr2.do`. Here is the file in printed form. Just skim it to begin.

```
─────────────────────────────────────────────── test_lr2.do ───────────
 1:  // version number intentionally omitted
 2:
 3:  clear all
 4:  run lr2.mata
 5:
 6:  sysuse auto
 7:  keep mpg weight displacement
 8:
 9:  // ------------------------------------------------------------
10:  // Utility routines to test "equality".
11:  mata:
12:  void virtuallyequal(real matrix A, real matrix B)
13:  {
14:          almostequal(A, B, 1e-15)
15:  }
16:
17:  void almostequal(real matrix A, real matrix B, real scalar tol)
18:  {
19:          assert(mreldif(A, B) <= tol)
20:  }
21:  end
22:
23:  // Prove that almostequal() works:
24:  mata:
25:  b1 = (1, 1, 1)
26:  b2 = (1, 1, 1+1e-14)
27:  end
28:  capture noisily mata: virtuallyequal(b1, b2)
29:  assert _rc!=0
30:
31:  // ------------------------------------------------------------
32:  // Test 1:  Model with an intercept.
33:
34:  regress mpg weight displacement
35:  predict double yhat
36:
37:  mata:
38:  st_view(X = ., ., ("weight", "displacement"))
39:  st_view(y = ., ., "mpg")
```

```
40: r = lrset(y, X, 1)
41: virtuallyequal(lr_b(r)´,    st_matrix("e(b)"))
42: virtuallyequal(lr_ee(r),    st_numscalar("e(rss)"))
43: virtuallyequal(lr_s2(r),    st_numscalar("e(rmse)")^2)
44: virtuallyequal(lr_V(r),     st_matrix("e(V)"))
45: virtuallyequal(lr_yhat(r), st_data(., "yhat"))
46: end
47:
48: // ------------------------------------------------------------
49: // Test 2:  Model without an intercept.
50:
51: reg mpg weight displacement, nocons
52: drop yhat
53: predict double yhat
54: mata:
55: st_view(X = ., ., ("weight", "displacement"))
56: st_view(y = ., ., "mpg")
57: r = lrset(y, X, 1)
58: virtuallyequal(lr_b(r)´,    st_matrix("e(b)"))
59: virtuallyequal(lr_ee(r),    st_numscalar("e(rss)"))
60: virtuallyequal(lr_s2(r),    st_numscalar("e(rmse)")^2)
61: virtuallyequal(lr_V(r),     st_matrix("e(V)"))
62: almostequal(   lr_yhat(r), st_data(., "yhat"), 1e-13)
63: end
64:
65: // ------------------------------------------------------------
66: // Test 3:  Zero RHS vars, with intercept.
67:
68: reg mpg
69: drop yhat
70: predict double yhat
71:
72: mata:
73: st_view(y = ., ., "mpg")
74: X = J(rows(y), 0, 0)                    // no RHS vars
75: r = lrset(y, X, 1)
76: virtuallyequal(lr_b(r)´,    st_matrix("e(b)"))
77: virtuallyequal(lr_ee(r),    st_numscalar("e(rss)"))
78: virtuallyequal(lr_s2(r),    st_numscalar("e(rmse)")^2)
79: virtuallyequal(lr_V(r),     st_matrix("e(V)"))
80: virtuallyequal(lr_yhat(r), st_data(., "yhat"))
81: end
82:
83: // ------------------------------------------------------------
84: // Test 4:  Zero RHS vars, without an intercept.
85:
86: reg mpg,  nocons
87: drop yhat
88: predict double yhat
89:
90: mata:
91: st_view(y = ., ., "mpg")
92: X = J(rows(y), 0, 0)                    // no RHS vars
93:
94: r = lrset(y, X, 0)
95: virtuallyequal(lr_b(r)´,    st_matrix("e(b)"))
96: virtuallyequal(lr_ee(r),    st_numscalar("e(rss)"))
97: virtuallyequal(lr_s2(r),    st_numscalar("e(rmse)")^2)
98: virtuallyequal(lr_V(r),     st_matrix("e(V)"))
```

```
 99:   virtuallyequal(lr_yhat(r), st_data(., "yhat"))
100:   end
101:
102:   // ------------------------------------------------------------
103:   // Test 5:  Collinear RHS var, with intercept.
104:
105:   reg mpg weight displacement weight
106:   drop yhat
107:   predict double yhat
108:
109:   mata:
110:   st_view(X = ., ., ("weight", "displacement", "weight"))
111:   st_view(y = ., ., "mpg")
112:
113:   r = lrset(y, X, 1)
114:   virtuallyequal(lr_b(r)´,    st_matrix("e(b)"))
115:   virtuallyequal(lr_ee(r),    st_numscalar("e(rss)"))
116:   virtuallyequal(lr_s2(r),    st_numscalar("e(rmse)")^2)
117:   virtuallyequal(lr_V(r),     st_matrix("e(V)"))
118:   virtuallyequal(lr_yhat(r), st_data(., "yhat"))
119:   end
120:
121:   // ------------------------------------------------------------
122:   // Test 6:  Collinear RHS var, without an intercept.
123:
124:   reg mpg weight displacement weight, nocons
125:   drop yhat
126:   predict double yhat
127:
128:   mata:
129:   st_view(X = ., ., ("weight", "displacement", "weight"))
130:   st_view(y = ., ., "mpg")
131:
132:   r = lrset(y, X, 0)
133:   virtuallyequal(lr_b(r)´,    st_matrix("e(b)"))
134:   virtuallyequal(lr_ee(r),    st_numscalar("e(rss)"))
135:   virtuallyequal(lr_s2(r),    st_numscalar("e(rmse)")^2)
136:   virtuallyequal(lr_V(r),     st_matrix("e(V)"))
137:   almostequal(  lr_yhat(r), st_data(., "yhat"), 1e-13)
138:   end
139:
140:   // ------------------------------------------------------------
141:   // Test 7:  Zero observations, with intercept.
142:
143:   mata:
144:   X = J(0, 3, .)
145:   y = J(0, 1, .)
146:
147:   r = lrset(y, X, 1)
148:   assert(lr_b(r)´ == J(1, 4, .))
149:   assert(lr_ee(r) == 0)
150:   assert(lr_s2(r) == 0)
151:   assert(lr_V(r)  == (J(3,3,0), J(3,1,.) \ J(1,4,.)))   // yuck!
152:   assert(rows(lr_yhat(r))==0 & cols(lr_yhat(r))==1)
153:   end
154:
```

```
155:  // ------------------------------------------------------------
156:  // Test 8:  Zero observations, without an intercept.
157:
158:  mata:
159:  X = J(0, 3,  .)
160:  y = J(0, 1,  .)
161:
162:  r = lrset(y, X, 0)
163:
164:  // Coefficients.
165:  assert(lr_b(r)´  == (0,0,0))
166:  assert(lr_ee(r) == 0)
167:  assert(lr_s2(r) == .)
168:  assert(lr_V(r)   ==   J(3,3,.))
169:  assert(rows(lr_yhat(r))==0 & cols(lr_yhat(r))==1)
170:  end
```
                                              ─────────── test_lr2.do ───────

Ignoring the line with the comment "yuck!" at the end, this is a lovely certification
script. It is more organized and readable than my certification scripts usually are.
More importantly, the script tests reasonable cases and extreme ones. Notice that I
tested `lr*()` with no covariates (independent variables), no intercept, no covariates or
intercept, and no observations, and I should have tested it with one and two observations
as well.

One reason `test_lr2.do` is so lovely is that I wrote a Mata function `virtuallyequal()`
so that I could write single lines such as

```
    virtuallyequal(lr_b(r)´,   st_matrix("e(b)"))
```

In my first draft, the test was more wordy:

```
    trub = st_matrix("e(b)")
    b = lr_b(r)
    b´   trub
    b´-trub
    mreldif(b´, trub)
    assert(mreldif(b´, trub) <= 1e-15)
```

There is something to be said for both approaches. The more wordy version displayed
results, and those results helped me to debug code when results were wrong. The extra
information also allowed me to choose the proper tolerances for the `assert()` statement.

When I was done, all the tolerances except two were 1e–15. The two exceptions were
both 1e–13 (lines 62 and 137 of `test_lr2.do`), and both involved predicted values for
models without intercept. I was not bothered by that because I know that Stata's
`regress` command calculates predicted values in quad precision, something I had not
bothered to do in the Mata code. Even if that had not been the explanation, a tolerance
of 1e–13 would not have concerned me. It is rare when I can write test scripts containing
tolerances of 1e–15. Tolerances of 1e–12 and 1e–13 are quite respectable.

I should explain about using tolerances like 1e–13 and 1e–15 instead of 1.0c–2c and 1.0x–32. I did, after all, lecture you in section 5.2.1.2 on the benefits of small values with exact and simple binary representations. My best defense, if only it were true, would be that I knew it did not matter in this case. Well, I do know it does not matter, but that was why I used 1e–13 and 1e–15. For the record, when all you want to do is verify that differences are small enough, it does not matter how you specify the tolerance.

My second best defense, if it were true, would be that I did not want to confuse you with statements like

```
assert(mreldif(b´, trub) <= 1.0x-32)
```

It is true that I do not want to confuse you, but I am perfectly willing to confuse you if it is for your own good. The truth is that I am sloppier when I write test scripts than when I write code, and I think in base 10 just like you do.

I seldom polish my test scripts, but I polished this one so that you would not have to struggle through the twists and turns that I introduced as I debugged **lr*()**. There are places in the original where, before an **assert()** statement that was failing, I displayed **lr_XXinv()** and even **invsym(lr_XXinv())** as I was backing through the steps to understand where the calculation went wrong. There would have been nothing wrong with leaving in the twists and turns. Test scripts need to test various cases and assert that results are accurate enough, but if they do more than that, it does not matter.

Finally, there is the story behind the "yuck!" in the line

```
assert(lr_V(r)  == (J(3,3,0), J(3,1,.) \ J(1,4,.)))  // yuck!
```

When there are zero observations—when X is $0 \times k$—**lr_V(r)** returns the inelegant result

```
: lr_V(r)
[symmetric]
          1    2    3    4

    1     0
    2     0    0
    3     0    0    0
    4     .    .    .    .
```

Notice the odd mix of 0s and missing values. Yuck.

This result is, I suppose, acceptable, but it would be better if **lr_V(r)** returned a matrix of all 0s, and it would be easy to fix the code. The problem traces back to the **lr*()** routine **calc_SSinv_cons()** where I add a row and column to **r.SSinv** to handle the inclusion of an intercept. If **lr_N(r)==0**, I need to append 0s instead. I could have fixed the code in less time than I have already spent telling you about it. I left the inelegance in the code for two reasons.

First, I want to tell you about extreme cases such as zero observations and zero variables. When I write code, I ignore the existence of such cases. I then include extreme cases in the certification script. I test the cases, and I usually observe satisfactory results just because of how Mata handles null matrices. Sometimes I observe an inelegant result as I do here, and then I decide whether the result is acceptable or I need to go back and fix it. I would have fixed this result. I did not because you might not have fixed it, and I wanted to discuss that with you, too.

Let's imagine that you wrote lr*(), discovered the embarrassing result, but decided it was acceptable. Because you are embarrassed about the result, you will be tempted to omit certifying it. Rather than assert that lr_V(r) is (J(3,3,0), J(3,1,.) \ J(1,4,.)), you will omit the line altogether. Someday, you think to yourself, you will go back and improve this. "I really should not include results that are not perfect in my script," you tell yourself.

If you are too embarrassed to include the line

```
    assert(lr_V(r)   ==  (J(3,3,0),  J(3,1,.) \ J(1,4,.)))
```

then you should fix the code. Whether you fix it or not, there are good reasons for including the **assert()**, whatever the values are. One purpose of certification scripts is to detect future changes in the behavior of your code. Code can change behavior even when it has not been modified. Results can change because of a change in an external routine that your code is calling. When results do change, that might be because the new result is better, or there has been a change in how the routine is to be used, or the routine has a newly introduced bug. The purpose of certification is to detect changes and stop. It is your job to determine why results changed and take the appropriate action.

11.5.2 Adding lr*() to the lmatabook.mlib library

Copy the following files to your approved source directory:

```
    lr2.mata
    test_lr2.do
```

Change to your approved source directory. Add the line "do test_lr2" to test.do:

```
─────────────────────────────────── test.do ───────────
// version number intentionally omitted
do test_n_choose_k
do test_lr2
─────────────────────────────────── test.do ───────────
```

Run it by typing do test.

Add the line "do lr2.mata" to make_lmatabook.do:

```
────────────────────────────────── make_lmatabook.do ─────────
// version number intentionally omitted
clear all
do hello.mata
do n_choose_k.mata
do lr2.mata
lmbuild lmatabook.mlib, replace
────────────────────────────────── make_lmatabook.do ─────────
```

Run it by typing do make_lmatabook.

Rerun test.do.

You are done.

12 Mata's classes

12.1 Overview

In this chapter, we cover classes as a programming concept. In the next chapter, we will reimplement `lr*()` as a class.

Classes are like structures, but with added features. You declare them in the same way that you declare structures, substituting the word `class` for `struct`:

```
class Foo
{
        member declarations
}
```

12.1.1 Classes contain member variables

Classes, like structures, can contain member variables:

```
class Foo
{
        real colvector    y
        real matrix       X
        real matrix       XX
        real colvector    Xy
}
```

This class is just like its corresponding structure would be. You refer to the class's member variables just as you would refer to them were `Foo` a structure. If you have declared

```
class Foo scalar    r
```

then `r` is an instance of class `Foo`, and you refer to its members as `r.y`, `r.X`, `r.XX`, and `r.Xy`. Variable `r` shares all the features of a structure variable. You can pass `r` as an argument to functions, you can write functions that return `r`, you can copy `r` to other variables via assignment, and you can test whether `r` equals `x` or does not equal `x` regardless of the kind of variable `x` is.

12.1.2 Classes contain member functions

Classes diverge from structures in that they can also contain member functions. The following class contains functions `setup()`, `calc()`, and `b()` along with the member variables `y`, `X`, etc.

```
class Foo
{
        real colvector    y
        real matrix       X
        real matrix       XX
        real colvector    Xy

        void              setup()
        void              calc()
        real colvector    b()
}
```

Member functions are defined in the same way as regular functions, but with added syntax to indicate their class membership. You add `Foo::` in front of the function's name:

```
void Foo::setup(real colvector user_y, real matrix user_X)
{
        . . .
}
void Foo::calc()
{
        . . .
}
real colvector Foo::b()
{
        . . .
}
```

There are two types of programmers where classes are concerned. There are insiders, who write the class's definition and its member programs, and outsiders, who use class's member variables and functions in the other programs they write. These two types of programmers could be the same person, but it is important which hat he or she is wearing because each type uses different notation to refer to the class's members.

Outsiders use the same notation as they would with structures. `r.y` and `r.b()` is how they refer to member variable `y` and member function `b()`:

```
... userfunction(...)
{
        class Foo scalar    r
        real colvector      b
        .
        .
        r.setup(y, X)
        .
        .
        ... r.y ... r.XX ...
        .
        .
        b = r.b()
        .
        .
}
```

Insiders omit the **r** prefix inside the bodies of member functions:

```
void Foo::setup(real colvector user_y, real matrix user_X)
{
        y = user_y             // y is the class's y
        X = user_X             // and so is X
        calc()                 // and so is calc()
}
void Foo::calc()
{
        XX = X'X               // XX and X are the class's XX and X
        Xy = X'y               // Xy and y are the class's Xy and y
}
real colvector Foo::b()
{
        return(invsym(XX) * Xy)
}
```

Insiders code y, X, and calc(). They do not code r.y, r.X, or r.calc().

12.1.3   Member functions occult external functions

Insiders can choose whatever names they wish for members, even if that name is already being used by Mata in some other context. For instance, Mata provides a built-in function named invsym(), but that does not prevent classes from providing their own function with the same name:

```
class Foo
{
        real colvector    y
        real matrix       X
        real matrix       XX
        real colvector    Xy

        void              setup()
        real colvector    b()
        void              calc()
        real matrix       invsym()          // <-- new
}
```

Including invsym() in the class definition changes the meaning of invsym() in Foo::b()'s code, which was and still is

```
real colvector Foo::b()
{
        return(invsym(XX) * Xy)
}
```

Previously, b() called Mata's built-in invsym() function that inverted symmetric matrices. Now b() calls the class's invsym() function. If b() still needed to call Mata's invsym(), the code would need to be changed to read

```
real colvector Foo::b()
{
        return(::invsym(XX) * Xy)
}
```

`::invsym()` specifies that the external `invsym()` is to be called, not the class's. Even when there is no class-provided function, programmers can code double colons in front of a function name if they want to emphasize that the function being called is defined outside of the class, but few programmers do that.

I added `invsym()` to the class just so that I could explain what would happen in the case of name conflicts. In the examples that follow, pretend I never added it.

12.1.4 Members—variables and functions—can be private

Another feature unique to classes is that members can be private. In class `Foo`,

```
class Foo
{
        real colvector    y
        real matrix       X
        real matrix       XX
        real colvector    Xy

        void              setup()
        real colvector    b()
        void              calc()
}
```

all members are public, which is the default. A better version of `Foo` would make only some of the members public and make the rest private:

```
class Foo
{
        private real colvector    y
        private real matrix       X
        private real matrix       XX
        private real colvector    Xy

        public  void              setup()
        private void              calc()
        public  real colvector    b()
}
```

Public members are just what you would imagine them to be. They are members that anyone may call or use. Public members can be called and used by insiders when they write `setup()`, `calc()`, and `b()`, and they can be called and used by outsiders who define a `class Foo scalar` in their programs.

Private members cannot be called or used by outsiders. It is not really programmers who are insiders or outsiders as I have claimed; it is the functions they write that are inside or outside of the class definition. Only inside functions—other class functions—

may access private members. Outsiders—programmers who define `class Foo scalar r` in their programs—will be prohibited from coding `r.y`, `r.X`, `r.XX`, `r.Xy`, and `r.calc()` in the programs they write. They may code `r.setup()` and `r.b()`, but if they call or access any of the other members of the class, Mata will issue an error:

```
: void myfunc(real colvector y, real matrix X)
> {
>            class Foo scalar   r
>
>            r.X = X
X not found in class Foo
r(3000);
```

It might be better if the error message read, "X is private in class **Foo** and therefore you cannot use it", but as far as Mata is concerned, `r.X` simply does not exist when compiling an outside function. Outside functions cannot access `r.X`, and neither can they access the private functions:

```
: void myfunc(real colvector y, real matrix X)
> {
>            class Foo scalar   r
>
>            r.setup(y, X)
>            r.calc()
function calc() not declared in class Foo
r(3000);
```

Making members private is useful. Think of the `lr*()` system that we wrote in the previous chapter. We could not block `lr*()`'s users from accessing the structure's internal variables; we just hoped they did not. We hoped they would call `lr_b(r)` instead of accessing `r.b`. It would have been better if we could have blocked them from accessing `r.b` altogether. With classes, we can.

In the class **Foo** definition that I showed you, I specified the privacy setting member by member. You can instead specify the setting for groups of members, and the result is far more readable:

```
class Foo {
    public:
        void                setup()
        real colvector      b()

    private:
        real colvector      y
        real matrix         X
        real matrix         XX
        real colvector      Xy
        void                calc()
}
```

12.1.5   Classes can inherit from other classes

Class aficionados will tell you that inheritance is the best feature of classes. It would be difficult for me to agree with that because most class programs that I write do not use inheritance. Yet I agree with the sentiment. Inheritance is spectacular when you need it.

Let's continue with class `Foo`, which, in case you could not discern it from the names of its members, concerns the calculation of solutions to linear regressions. Let's pretend that `Foo` has been fleshed out with more features and is up, running, and debugged. It is a wonderful linear-regression engine. It is so good that it is no longer named `Foo`. It has been renamed `LinReg`.

```
class LinReg {
    public:
        void            setup()
        real colvector  b()
        real scalar     Rsquared()

    private:
        real colvector  y
        real matrix     X
        real matrix     XX
        real colvector  Xy
        void            calc()
}
```

ANOVA, another statistical estimator, is a special case of linear regression in that we need to make many of the same calculations, while other calculations need to be made differently or are unique to ANOVA itself. One way of programming ANOVA would be to inherit the parts in common from `LinReg`. Another way of saying the same thing is that we could extend the linear-regression class. We could do that by coding

```
class Anova extends LinReg
{
    public:
        void            setup()
        real scalar     model_SS()
        real vector     factor_SS()
        real scalar     resid_SS()
        real scalar     effects()

    private:
            .
            .
            .
}
```

The short phrase "`Anova extends LinReg`" causes `Anova` to inherit the features of `LinReg`. In writing `Anova`, we can use some of `LinReg`'s features, override others, and add new ones. Meanwhile, users of `Anova` will remain unaware that we built it from `LinReg`. We might tell them about the relationship if `LinReg` had a great reputation for accuracy and robustness, but other than for marketing purposes, our use of `LinReg` would be of no importance to them. `Anova` is a class, and to users, it is merely an implementation detail that it was built on top of `LinReg`.

One effect of `Anova extends LinReg` is that the functions and variables of `LinReg` become part of `Anova`. Think of it like this:

```
class Anova   =        class LinReg


                       class Anova additions

```

Here is a more detailed view:

```
class Anova   =        class LinReg
                          setup()
                          b()
                          Rsquared()
                          y
                          X
                          XX
                          Xy
                          calc()
                          .
                          .

                       class Anova additions
                          setup()
                          model_SS()
                          factor_SS()
                          resid_SS()
                          effects()
                          .
                          .
```

Notice that function `setup()` appears twice in class `ANOVA`. We will get to that later in section 12.1.5.2. In the meantime, note that `Rsquared()` is in `LinReg`'s contribution to `Anova` and `effects()` is in the additions. That will make no difference to users who have declared

```
    class Anova scalar   A
```

They will be able to call `A.Rsquared()` just as they can call `A.effects()`, yet all we will have to write is `Anova::effects()`! Even better, when we write `effects()`, we can even use `LinReg`'s private functions should that be convenient.

12.1.5.1   Privacy versus protection

Privacy comes in two forms where inheritance is concerned: `private` and `protected`. When I wrote that class members can be public or private, I should have written that they can be `public`, `protected`, or `private`.

`private` is super private. Only the declaring class can access the `private` members. Everyone else is blocked, and that includes inheritors.

When I wrote that `A.effects()` can use `LinReg`'s private functions, that was true only if you paid attention to typeface. I should have written that `A.effects()` can use `LinReg`'s `protected` functions and variables. `protected` members can be accessed by inheritors.

The usual terminology for the bequestor and the inheritor is superclass and subclass. In this terminology, `LinReg` is the superclass and `Anova` is the subclass. Another popular terminology refers to `LinReg` as the parent and `Anova` as the child.

Whatever we call `LinReg`—superclass, parent, or rich-in-features uncle—we will need to modify the class to expose its private members to inheritors:

```
class LinReg {
    public:
        void                    setup()
        real colvector          b()
        real scalar             Rsquared()
    protected:
        real colvector          y
        real matrix             X
        real matrix             XX
        real colvector          Xy
        void                    calc()
}
```

In a more fleshed out example, `LinReg` would have more members, and we would consider which should be `private` and which should be `protected`.

12.1.5.2   Subclass functions occult superclass functions

In the story I am telling, we are splicing `Anova` onto `LinReg` after the fact, but that is not how the situation usually unfolds. Programmers writing superclasses usually know that they are writing for inheritors. Sometimes they are writing the class for two audiences, as in the `LinReg` example. The class by itself is useful for solving linear-regression problems, and the class is also useful to programmers who wish to write other systems by extending it. Other times, the superclass is written only to be a superclass, and subclass extensions are necessary to make the combined object do something useful.

In the example I have shown you, function `A.Rsquared()` is provided by the superclass, and `A.effects()` is provided by the subclass. Functions can also be defined by both classes.

Let's imagine a function `Gstat()` that is defined by both `LinReg` and `Anova`. If a user of `Anova` calls `A.Gstat()`, it will be `Anova`'s function that is invoked. The jargon I like for this is that `Anova::Gstat()` *occults* `LinReg::Gstat()`, but the jargon used is that `LinReg::Gstat()` is shadowed by `Anova::Gstat()`. Whatever we call the behavior, it is useful for dealing with situations that must be handled differently. Let's assume that `Gstat()` calculates G statistics, whatever they might be. If G statistics are calculated one way for linear regression and another way for ANOVA, the programmer of `Anova` simply defines a new `Anova::Gstat()` function that effectively replaces `LinReg`'s `Gstat()` function.

`Anova::Gstat()` can even call `LinReg::Gstat()` if necessary. Perhaps ANOVA's G statistic is linear regression's G statistic multiplied by $k/(k-1)$. In that case, the code for `Anova::Gstat()` could read

```
real scalar Anova::Gstat()
{
        return( super.Gstat()*(k/(k-1)) )
}
```

Another example of a function provided by both `LinReg` and `Anova` is `setup()`. Both classes already declare the function. Users of either class will call `setup()` to specify the linear-regression or ANOVA problem to be fit. `Anova`'s `setup()` function might be

```
void Anova::setup(real colvector y, real matrix Factors)
{
        real matrix    X
        .
        .
        X = ... Factors ...
        super.setup(y, X)
        .
        .
}
```

The `super` prefix may be used with functions or variables.

12.1.5.3   Multiple inheritance

Once `Anova` is written, we could write another class that inherits from it! One-way ANOVA is a special case of ANOVA. We could write class `OneWay` extends `Anova`.

```
class OneWay extends Anova
{
        .
        .
        .
}
```

`OneWay` will doubtlessly have its own `setup()` function. If `OneWay`'s `setup()` needs to call `Anova`'s `setup()` function, it would call `super.setup()`. If it needs to call `LinReg`'s

setup(), it would call super.super.setup(). OneWay's setup() is unlikely to need
to do that, however, because Anova's setup() should handle the issue of initializing
LinReg.

12.1.5.4   And more

We will discuss Mata's other inheritance features in section 12.6. Mata provides virtual
and final functions and supports polymorphisms.

12.2   Class creation and deletion

The creation and deletion of class instances is automatic in Mata. The situation here
is the same as it was for structures; we mentioned this in passing in section 10.8.

First, variables are automatically filled in with class instances if you remember to declare
them as scalars. If you forget, the variable will default to being a matrix, which means
it will be 0×0 initially.

Second, just as with structures, Mata automatically creates a corresponding constructor
function when you define a class. If the class is named Foo, its constructor function is
Foo(). Foo() without arguments returns a new 1×1 class instance. Foo(r) returns r
new instances in a $1 \times r$ vector. Foo(r, c) returns rc new instances in an $r \times c$ matrix.
If you declare

```
        class Foo vector    vf
```

then vf starts as 1×0. If vf needs to contain four class instances, you can subsequently
code

```
        vf = Foo(4)
```

If you were to declare

```
        pointer(class Foo scalar) scalar   p
```

and p needed to point to a new class instance, you can subsequently code

```
        p = &(Foo())
```

Classes have two other related features. If you have class functions named new() and
destroy(), they will be run automatically at the appropriate times.

When member function Foo::new() exists, it is run whenever a new instance of Foo is
created. Foo::new() is run when you use the Foo() function, and it is run when you
declare Foo variables to be scalars. The running of Foo::new() becomes part of Mata's
instance-creation process for Foo. You might define a new() that reads

```
void Foo::new()
{
        a = b = .z
        name  = "default"
}
```

If you did, every new instance of Foo will have a and b set to .z and name set to "default".

At the opposite end of the process is destroy(). If Foo::destroy() exists, it will be run when instances of Foo are deleted. Just as with new(), you do not call destroy(). The running of Foo::destroy() becomes part of Mata's instance-deletion process for Foo. You might define a destroy() that reads

```
void Foo::destroy()
{
        if (filehandle != .) fclose(filehandle)
}
```

This destroy() function would ensure that a file would be closed when member variable filehandle had been filled in with a nonmissing value.

There are situations in which new() and destroy() are useful, but they do not arise as often as in other languages because memory management in Mata is automatic. C and C++ users: It is not your responsibility to allocate memory for class instances, nor is it your responsibility to free it.

destroy() is needed only for special cases, such as closing an open file.

new() is sometimes useful, but it is often unnecessary because Mata's automatic preinitialization is sufficient. In most code that I write, I am forced by the nature of the problem to include a setup function that users must call anyway. Remember the lrset() function of the previous chapter? Users had to call it to specify the model to be fit. When you have such a function, it is easier to place any initialization code in that function than to put the code in a separately defined new() function that is automatically called. The result will be the same either way.

Although memory management in Mata is automatic, you can hurry the process. Say you have a class Foo scalar foo that, late in your program, is no longer needed even though the program still has a few more things to do. You can reduce foo's memory footprint by coding

```
foo = Foo()
```

If you have a class vector fv, code

```
fv = J(1, 0, .)
```

If you have a class matrix fm, code

```
fm = J(0, 0, .)
```

If you have a pointer **p** to an instance or instances of **Foo**, code

```
p = NULL
```

None of this is necessary. The memory of class instances is freed when the function declaring them exits or returns, the exception being any class instances returned by the function. Those instances then become the property of the caller, and the caller frees them when it returns unless it returns them, and so on.

12.3 The this prefix

When writing member functions, you may refer to member variables and functions directly. If y, X, and **init()** are members of **Foo**, you may refer to them in just that way when writing member functions:

```
void Foo::setup(real colvector user_y, real matrix user_X)
{
        y = user_y              // y is Foo´s y
        X = user_X              // X is Foo´s X
        init()                  // init() is Foo´s init()
}
```

You can emphasize that y, X, and **init()** are members of the class by prefixing them with **this** if you wish:

```
void Foo::setup(real colvector user_y, real matrix user_X)
{
        this.y = user_y
        this.X = user_X
        this.init()
}
```

You need not be consistent. You could prefix y but not X, or prefix X but not **init()**.

Use of **this** is sometimes required, as in

```
void Foo::setup(real colvector y, real matrix X)
{
        this.y = y
        this.X = X
        init()
}
```

References to y and X in the above code are references to the arguments. You code **this.y** and **this.X** to specify Foo's member variables. Specifying **this.y** and **this.X** would also have been required if you declared y or X to be variables inside the function.

Whether you use **this** or not, member functions always occult external functions. It makes no difference whether you code **init()** or **this.init()**. If you later want to call external function **init()**—**init()** outside of the class—you must code **::init()**.

12.4   Should all member variables be private?

Some would argue—me among them—that all member variables should be `private`. If
users of the class need to access them, that should be done using access functions. That
is, rather than define

```
class Foo
{
        .
        .
        real matrix     X
        .
        .
}
```

you should define

```
class Foo
{
        .
        .
        private real matrix     X
        real matrix             X()
        .
        .
}

real matrix Foo::X()
{
        return(X)
}
```

There are two advantages to the variable-plus-function approach. The first is that users
cannot change the contents of X. The second is that you, the programmer of the class,
now have the ultimate freedom to modify the code. Today you are storing X, but
someday you may be storing Xdev, X as deviations from the means. Should that day
arrive, you can keep old code working by modifying the function X() to read

```
real matrix Foo::X()
{
        return(Xdev :+ Xmeans)
}
```

If you planned all along that users set X, provide a function to do that, too:

```
void Foo::setX(real matrix X)   this.X = X
```

Now that you have a **setX()** function, you can add code to it to verify that the matrix that users specify is as you expect it to be:

```
void Foo::setX(real matrix X)
{
        if (!issymmetric(X)) {
                _error(3300, "symmetric matrix required")
        }
        this.X = X
}
```

I often combine the two functions into one:

```
/*
        _
        X = foo()                     return current value of X

        (void) foo(X)                 set value of X
                _
*/

transmorphic Foo::X(|real matrix userX)
{
        if (args==0) return(X)

        if (!issymmetric(X)) {
                _error(3300, "symmetric matrix required")
        }
        this.X = X
}
```

If your response is that you do not need to do any of this, then X should have been private from the outset. If you make variables public, you must assume that someone, someday will access them and even change their contents.

12.5   Classes with no member variables

Classes are not required to contain variables, and classes that contain only functions can be surprisingly useful. Here is one that provides just one member function:

```
/*
    _____

    miles = e.distance((latitude, longitude), (latitude, longitude))
                       --------  ---------    --------  ---------

        where -e- is a -class EarthDistance scalar-
*/
class EarthDistance
{
        real scalar distance()
}
real scalar EarthDistance::distance(real vector pos1,
                                    real vector pos2)
{
            .
            .
}
```

I will eventually show you the code. The one function that this class provides returns the distance in miles between two positions on the Earth expressed in degrees of latitude and longitude. I can use the function interactively,

```
: e = EarthDistance()
: e.distance( (39, -77), (40, -78) )
  87.31935772
```

or in a program,

```
... myfcn(...)
{
        class EarthDistance scalar    e
        .
        .
        s = e.distance( (39, -77), (40, -78) )
        .
        .
}
```

You must be asking yourself why anybody would create a class to provide just one function. There are two reasons. First, if the calculation is complicated and requires subroutines, classes provide a way to encapsulate the related code and so keep the subroutines private. Second, you can give the functions and subroutines better names because they have a context.

Naming a function `distance()` will not do. The name is too general. Distance between what? I have written lots of distance functions, including one that calculated the distance between matrices. Even putting that argument aside, there are reasons we should not name the function `distance()`. Standalone functions share a common name space. That name space is shared by you, me, everybody else, and, oh yes, StataCorp. Is somebody else using the name `distance()`? Probably. Even if we do not care about that, we need to worry that StataCorp might someday take the name. If they do, their function will occult ours, and that will inconvenience us.

Classes circumvent these problems. It is the class's name that needs to be unique, not the function's name within the class. We could choose a name even more ungainly than `EarthDistance` and we would not care much because we only have to specify the class's name on the declaration statement. Look at the previous example. We used the word `EarthDistance` once, in the declaration of the class scalar, and after that we got to use the nice name `e.distance()` in the code.

Here is the code for `EarthDistance`. You can see that I defined three private subroutines—`radians()`, `hav()`, and `invhav()`—which made it easier for me to write the one public function `distance()`.

```
                                                       ── earthdistance.mata ───────
  1:  *! version 1.0.0
  2:
  3:  version 15
  4:  set matastrict on
  5:
  6:  mata:
  7:
  8:  class EarthDistance
  9:  {
 10:      public:
 11:          real scalar  distance()
 12:
 13:      private:
 14:          real scalar  radians()      // radians from degrees
 15:          real scalar  hav()          // haversine function
 16:          real scalar  invhav()       // inverse haversine function
 17:  }
 18:
 19:  real scalar EarthDistance::distance(real vector pos1,
 20:                                      real vector pos2)
 21:  {
 22:          real scalar  dlon, dlat
 23:          real scalar  lat1, lat2
 24:          real scalar  lon1, lon2
 25:          real scalar  h, radius
 26:
 27:          radius = 3961                // radius of Earth, in miles
 28:
 29:          lat1 = radians(pos1[1])
 30:          lon1 = radians(pos1[2])
 31:
 32:          lat2 = radians(pos2[1])
 33:          lon2 = radians(pos2[2])
 34:
 35:          dlon = lon2 - lon1
 36:          dlat = lat2 - lat1
```

```
37:
38:              h = hav(dlat) + cos(lat1)*cos(lat2)*hav(dlon)
39:              if (h>1) return(.)
40:
41:              return(radius*invhav(h))
42:      }
43:
44:      real scalar EarthDistance::hav(real scalar theta)
45:                                  return((1-cos(theta))/2)
46:
47:      real scalar EarthDistance::invhav(real scalar h)
48:                                  return(2*asin(sqrt(h)))
49:
50:      real scalar EarthDistance::radians(real scalar d)
51:                                  return(d*(pi()/180))
52:
53:      end
```
———————————————————————————————————— earthdistance.mata ————

Note: The formulas above calculate the distance in miles between points on a sphere.
The Earth is not a sphere. To produce distances in kilometers, change `radius = 3961`
to `radius = 6375`.

Once you start writing function-only classes, you sometimes find reasons to add member
variables to them later. The code above produces distance in miles. We could modify
`distance()` to produce results in miles or kilometers by adding a setting. We could
do that by moving variable `radius` out of `distance()` and into the class (making it
private, of course) and adding a public `setup()` routine:

```
void EarthDistance::setup(string scalar units)
{
        if      (units=="miles")      radius = 3961
        else if (units=="kilometers") radius = 6375
        else    _error("miles or kilometers required")
}
```

We might now decide to also allow custom radiuses so that users could calculate distances on other planets. If we did that, we should also change the name of the class
`EarthDistance` to `PlanetaryDistance`.

12.6 Inheritance

This section continues where section 12.1.5 left off. We were discussing inheritance and
stopped before discussing inheritance's advanced features, namely, `virtual`, `final`, and
polymorphisms. Before continuing that discussion, however, we need to consider another
advanced topic. When should you use inheritance, and what are the alternatives to it?
We left off with an example of the class `Anova extends LinReg`:

```
class Anova extends LinReg
{
        .
        .
        .
}
```

In the example we are discussing, class `LinReg` provided function `Rsquared()` that users of the class `Anova` could use, yet we did not have to write it. The situation was as if `LinReg::Rsquared()` had been custom written for ANOVA problems. There is, however, another way we could have organized the code. We could have put class `LinReg` inside class `Anova`:

```
class Anova
{
        .
        .
        private class LinReg scalar    lr
        .
        .
}
```

If we had written the code this way, `Anova` users could not have called `LinReg`'s `Rsquared()`, but put that aside for a moment. If the other reason we wanted `Anova` to extend `LinReg` was merely so that we could call `LinReg` functions in the new `Anova` code we would write, then the class-in-a-class solution would have been a better way to organize our code than the inheritance solution would have been. Class-in-a-class is simpler and easier to understand.

I previously offered the example `Anova::setup()` when `Anova` extended `LinReg`. It was

```
void Anova::setup(real colvector y, real matrix Factors)
{
        real matrix    X
        .
        .
        X = ... Factors ...
        super.setup(y, X)
        .
        .
}
```

If `LinReg` was instead put inside `Anova`, only one line of the above code would need to change. Instead of calling `super.setup()`, we would call `lr.setup()`:

```
void Anova::setup(real colvector y, real matrix Factors)
{
        real matrix    X
        .
        .
        X = ... Factors ...
        lr.setup(y, X)
        .
        .
}
```

This example suggests that the `Anova` code would be no more difficult to write, but in fact, the code would be even easier to write using the class-in-a-class style because we would not have to write our new code to fit into the framework and style established by `LinReg`. We would just use its functions as we needed them. Yes, if we wanted to provide `Anova`'s users with an `Rsquared()` function, we would have had to write it, but it would not have been difficult. The code for the entire function fits on one line:

```
real scalar Anova::LinReg() return(lr.Rsquared())
```

If this makes it seem as if inheritance is a gratuitous complication instead of a useful programming tool, it is not. When I said that the class-in-a-class code would be easier to write because our new code would need to fit into the framework established by `LinReg`, that disadvantage can turn into an advantage and our new code would be easier to write if `LinReg` is well designed and ANOVA really is a conceptual extension of linear regression. You have to make a judgment. For instance, in no sense is `Anova` a conceptual extension of string utility functions that chop text strings, put them together again, and the like. That does not mean that string utilities might not be useful in implementing `Anova`; it just means that if you wanted to use a class `StringUtilities` that provided those functions, you would include the class inside of `Anova` and not inherit from it.

When inheritance is justified—when the new code you are writing really is a conceptual extension of an existing class—using the advanced features we are about to discuss will allow you to do things that would take pages and pages of complicated code if done another way.

12.6.1 Virtual functions

Assume that we want to create two different classes, one for fitting linear regression and another for fitting logistic regression. These two estimators are examples of something called linear models in statistics, and we intend to implement other linear models later. A good way to write that code would be to write a linear-model class. In this case, I know I want to use inheritance because linear-model statistics are conceptually related to the statistics of the individual linear models. For this example, however, I am going to focus on something more mundane, namely, the displaying of results after fitting a linear model. The linear-model class looks like this:

```
class LinearModel
{
        .
        .
        void          displayresults()
        .
        .
}
```

Class `LinearModel` does not by itself do anything useful. We will write `LinearModel` solely to provide useful subroutines, such as `displayresults()`, which will display the estimation results.

We will write the two individual estimation classes using inheritance. Subroutines for formulas in common will appear in the superclass `LinearModel`, and formulas that vary model by model will appear in the subclasses. `Rsquared()` is an example of the latter:

```
class LinearRegression extends LinearModel
{
        .
        .
        real scalar    Rsquared()
        .
        .
}
class LogisticRegression extends LinearModel
{
        .
        .
        real scalar    Rsquared()
        .
        .
}
```

The `Rsquared()` functions return a scalar value related to how well the models fit on a scale of 0 to 1. How you calculate the value differs depending on which model was fit, which means that the `Rsquared()` function is defined in the subclasses. The function will be called by various routines in the overall system, but I want to focus on it being called by `LinearModel::displayresults()`, the routine that displays the results. I imagine the system being used like this by someone fitting a linear regression:

```
class LinearRegression scalar    m
m.setup(y, X)
m.fit()
m.displayresults()
```

If the user needed to fit a logistic regression, m would be a `class LogisticRegression scalar`, but nothing else would change:

```
class LogisticRegression scalar    m

m.setup(y, X)
m.fit()
m.displayresults()
```

In both cases, users code `m.displayresults()` to see the final results:

```
m.displayresults()
   (output omitted)
```

Callers will code `m.displayresults()` but `LinearModel::displayresults()`. I previously described how to think about inherited classes. I drew the figures like this:

```
                            ┌────────────────────────────┐
                            │      class LinearModel      │  <-- superclass
LinearRegression    =       ├────────────────────────────┤
                            │   class LinearRegression    │  <-- subclass
                            └────────────────────────────┘

                            ┌────────────────────────────┐
                            │      class LinearModel      │  <-- superclass
LogisticRegression  =       ├────────────────────────────┤
                            │  class LogisticRegression   │  <-- subclass
                            └────────────────────────────┘
```

The way inheritance works, subclass functions can call superclass functions but not vice versa because, while the subclass knows about the superclass, the superclass knows nothing about the subclass. Inheritance is a one-way street heading north.

That will be a problem for the system I described because we need the superclass's `displayresults()` function to call `Rsquared()`, but `Rsquared()` is defined by the subclass. The solution is to add a virtual placeholder function named `Rsquared()` to the superclass:

```
class LinearModel
{
        .
        .
        void                     displayresults()
        virtual real scalar      Rsquared()
        .
        .
}
```

The virtual placeholder specifies that if `LinearModel::Rsquared()` is called, then `LinearModel` is to execute the function named `Rsquared()` in the subclass. You are not required to define the code for `LinearModel::Rsquared()` because `LinearModel`'s function will be called only if the subclass does not provide an `Rsquared()` function or if there is no subclass. `LinearModel::Rsquared()` is for emergency use only. I tend to define the function anyway and code

```
real scalar LinearModel::Rsquared() return(.)
```

It is easy to understand virtual when there are only two classes, one superclass and one subclass. Let me show you how it works when there are more subclasses. Consider the case where D extends C extends B extends A, which is figuratively

```
┌─────────────────────────────────┐
│           class A               │        <-- super-most class
├─────────────────────────────────┤
│           class B               │
├─────────────────────────────────┤
│           class C               │
├─────────────────────────────────┤
│           class D               │
└─────────────────────────────────┘
```

Now assume a function in class A calls f(), which is declared to be virtual in A. Which f() is called if more than one class defines it? Answer: The first f() not marked virtual or, if all are marked virtual, the bottommost f() defined. You search from top to bottom and skip any classes that do not define f().

Functions may be virtual, but virtual variables are not allowed.

12.6.2 Final functions

`final` is like `virtual` in that it too is set by the superclass:

```
class LinearModel
{
        .
        .
        void          displayresults()
        virtual real scalar  Rsquared()
          final real matrix  ingredient1()
        .
        .
}
```

Final means the opposite of virtual. Virtual means that the superclass calls the subclass's function. Final means that the subclass calls the superclass's function. A `final` function is the final word as far as inheritors are concerned.

Final member variables are allowed, too. If the superclass defined `basecalc()` and `N_of_obs` as final, then inheritors may not themselves define `basecalc()` or `N_of_obs`. All inheritors will use the superclass's function and variable.

Classes allow three kinds of members:

Kind of member	Meaning
regular (indicated by silence)	functions and variables
virtual	functions only
final	functions and variables

If the superclass defines member as...	and the subclass defines a member of the same name, then...
regular	subclass uses the subclass's member superclass uses the superclass's member
virtual	subclass uses the subclass's function superclass uses the subclass's function
final	superclass uses the superclass's member subclass uses the superclass's member subclass may not even define the member

Note: If the subclass does not define a member of the same name as the superclass, then whether the superclass's member is regular, virtual, or final, the superclass's member is used both by the superclass and by the subclass.

When you declare a function to be final, you are making an announcement: "Whatever it is that this function does, it is the final word on how to do it. Subclasses, if any, may not override this function. They may not occult this function. They may not even define a function of the same name lest anyone be misled."

Why would you want to do this? I showed an example above where `ingredient1()` was declared final in `LinearModel`. Let's imagine that `ingredient1()` makes a calculation of I_1 that, if it is calculated, must be calculated when the model is fit. I_1 is required for doing postestimation hypothesis testing, which will be performed by other functions that can be called after the model has been fit. Those functions combine I_1 with other calculations they make, but those calculations will only be mathematically valid if I_1 is calculated in the way they expect, which is how `ingredient1()` calculated it. By declaring `ingredient1()` to be final, the programmer of the superclass is ensuring that the calculation is made in the way that the postestimation functions expect.

If a class extending `LinearModel` attempted to define its own `ingredient1()` function, Mata would refuse to compile the code:

```
: class LinearRegression extends LinearModel
> {
>          .
>          .
>          real matrix ingredient1()
>          .
>          .
> }
ingredient1():  function declared final may not be modified
r(3000);
```

12.6.3   Polymorphisms

Assume that class B extends A and that variable b is declared to be a class B scalar. Obviously, b could be passed to functions expecting to receive a class B scalar. Surprisingly, b can also be passed to functions expecting to receive a class A scalar. b is a polymorph. It is both a class B and a class A scalar.

How can b be both? Let a be a class A scalar and b be a class B scalar. I drew pictures of inheritance, and you need to take them literally:

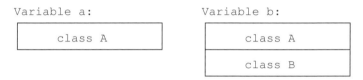

Variable b not only contains a class A object, the top part of b *is* a class A object.

Consider a program expecting a class A scalar, such as

```
... myprogram(class A scalar a, ...)
{
        .
        .
}
```

Because of polymorphism, `myprogram()` is allowed to receive a class B scalar. Now imagine that `myprogram()` calls `a.draw()`:

```
... myprogram(class A scalar a, ...)
{
        .
        .
        ... a.draw() ...
        .
        .
}
```

If myprogram() received a class B scalar, and if A::draw() is virtual, then myprogram()'s call to a.draw() will in fact execute B::draw()! This aspect of polymorphism can be put to good use. Imagine the following set of classes:

```
class Circle       extends   Object
class LineSegment  extends   Object
class Triangle     extends   LineSegment  extends  Object
class Rectangle    extends   LineSegment  extends  Object
```

The purpose of all these classes is to draw shapes on your monitor. Here is how they fit together to achieve that:

Class Object provides the lowest-level screen-drawing functions.

Class Circle draws circles by calling Object's screen-drawing functions.

Class LineSegment draws lines by calling Object's screen-drawing functions.

Class Triangle draws triangles by calling LineSegment to draw each of the legs.

Class Rectangle draws rectangles by calling LineSegment to draw each of the sides.

Here is a program for drawing a collection of objects:

```
void drawobjects(class Object vector o)
{
        real scalar   i

        for (i=1; i<=length(o); i++) o[i].draw()
}
```

The vector o that drawobjects() receives is not really a collection of pure class Objects. It is a collection of classes Circles, LineSegments, Triangles, and Rectangles. Indeed, o[1] might be one thing and o[2] another. That is allowed because the classes are polymorphisms of class Object.

The function will draw all the objects in the vector because draw() is virtual. Each class defines its own draw() function, and thus the call to a.draw() will sink to the bottommost draw() for each object.

12.6.4   When to use inheritance

Now you know why I said that inheritance can be spectacular when you need it. Yes, inheritance adds complication, but that complication turns into features in some cases. In cases where it does not, the complication is just added complication.

The reasons to use inheritance are as follows:

1. The subclass needs to export to its callers one or more of the public functions provided by the superclass.

2. The subclass needs to supply virtual functions for use by the superclass.

3. There will be multiple subclasses that exploit polymorphisms as I did in the `drawobjects()` routine above.

4. The subclass needs to call protected functions of the superclass.

12.7 Pointers to class instances

Pointers to class instances work just like pointers to structure instances, which we discussed in section 10.10.

Pointers to class member functions are not allowed. You may not code

```
p = & (myclass.myfcn ())
```

You may, however, code

```
p = &myclass
```

and subsequently call the function by coding

```
p->myfcn (...)
```

13 Programming example: Linear regression 2

13.1 Introduction

We wrote a first-rate linear-regression system using structures in chapter 11. We could have written it as a class, and that is exactly what we are going to do in this chapter. I will translate lr*() version 2 and produce the equivalent class **LinReg** code, and then I will tell you about it. For instance, function **lr_b()** for obtaining the linear-regression coefficients looked like this:

```
real colvector lr_b(struct lrinfo scalar r)
{
        if (r.b == .z) {
                if (r.cons) {
                        r.b          = lr_SSinv(r) * lr_St(r)
                        r.b[lr_K(r)] = lr_mean_y(r) -
                                            lr_mean_X(r)*lr_b_X(r)
                }
                else    r.b = lr_XXinv(r)  * lr_Xy(r)
        }
        return(r.b)
}
```

In the new code, it will look like this:

```
real colvector LinReg::b()
{
        if (b == .z) {
                if (cons) {
                        b      = SSinv() * St()
                        b[K()] = mean_y() - mean_X()*b_X()
                }
                else    b      = XXinv() * Xy()
        }
        return(b)
}
```

The new code looks simpler because there is less superfluous ink. The class provides a context, so what was **r.b[lr_K(r)]** becomes **b[K()]**. The same is true of all the other member functions, so a line that previously read

```
r.b[lr_K(r)] = lr_mean_y(r) - lr_mean_X(r)*lr_b_X(r)
```

will now read

```
b[K()] = mean_y() - mean_X()*b_X()
```

I will call the translated code **LinReg** version 1.

Then, because I want to show you how easy it is to add new features to class-based, self-threading code, we will modify the system to include two new standard-error calculations, the OPG-based variance matrix and the robust (sandwich) variance matrix. The new code, **LinReg** version 2, will be good enough to save in the **lmatabook.mlib** library, and we will do that.

Before we start, let me show how callers will use **LinReg**.

13.2 LinReg in use

LinReg works just like **lr*()** except for the differences in syntax because **LinReg** is a class, and those differences are all for the better. Whereas callers of **lr*()** coded

```
struct lrinfo scalar     r
r = lrsetup(y, X, 1)
lr_b(r)
lr_V(r)
```

callers of **LinReg** will code

```
class LinReg scalar     r
r.setup(y, X, 1)
r.b()
r.V()
```

Just as when I demonstrated the `lr*()` system, I have already defined y and X in Mata:

```
: y
              1

    1        22
    2        17
    3        22
    4        20
    5        15
    6        18
    7        26
    8        20
    9        16
   10        19
```

```
: X
              1        2

    1      2930      121
    2      3350      258
    3      2640      121
    4      3250      196
    5      4080      350
    6      3670      231
    7      2230      304
    8      3280      196
    9      3880      231
   10      3400      231
```

Here is `LinReg` used interactively:

```
: r = LinReg(1)
: r.setup(y, X, 1)   // set problem
: r.b()                    // coefficient vector
                    1

    1     -.0056644997
    2      .0004631627
    3      37.92487622
```

```
: r.V()                  // variance matrix
[symmetric]
                    1                2                3

    1     4.17119e-07
    2    -1.16730e-06      .0000252448
    3    -.0011030394     -.0018340879     4.120595787
```

```
: r.Vrobust()          // (will be added in LinReg version 2)
[symmetric]
                    1                2                3

     1     1.73849e-07
     2    -4.23393e-07     .0000160395
     3    -.0005087227    -.0020614122     2.309071791
```

```
: r.Vopg()             // (will be added in LinReg version 2)
[symmetric]
                    1                2                3

     1     1.38511e-06
     2    -4.26186e-06     .0000544914
     3    -.0034598706     .0012920538     10.88787184
```

```
: r.ee()               // error sum of squares
7.133114624
```

```
: r.s2()               // variance of e
1.019016375
```

```
: r.k()                // # of X vars, excluding intercept
2
```

```
: r.K()                // # of X vars, including intercept
3
```

```
: r.K_adj()            // r.K adj. for dropped (collinear) variables
3
```

```
: r.yhat()             // predicted values
                    1

     1     21.38393493
     2     19.06829838
     3     23.02663983
     4     19.60603225
     5      14.9758246
     6     17.24315309
     7     25.43384347
     8     19.43609726
     9     16.05360817
    10      18.772568
```

13.3   LinReg version-1 code

You can see a full listing of the **LinReg** version-1 code by typing in Stata

```
. view ~/matabook/linreg1.mata
```

If you have not yet downloaded the files associated with this book, see section 1.4.

The new code needs no explanation other than it is a line-by-line translation of the original structure-based code. Notice that member variables are now private, meaning that users of LinReg cannot access them directly, and that LinReg's internal subroutines are now private, too.

13.4 Adding OPG and robust variance estimates to LinReg

To show you how easy it is to add new features to self-threading code, I will now add two of them:

r.Vopg()   Variance based on the outer-product of gradients

r.Vrobust()   Variance based on the sandwich estimator

These two new features are alternatives to the existing likelihood-variance estimates provided when users call r.V().

I could have perhaps made my point by adding something easier, such as r.Rsquared(). I chose to add r.Vopg() and r.Vrobust() specifically because they would be difficult. Splicing alternative variance estimates into existing code would ordinarily involve more work than we will need to do.

The formulas for the new variance estimates are

―――――――――――――――――――――――― Formula sheet 3 ―――――――

```
V_opg = (s2)^2*((N-1)/N)*invsym(P)

V_robust = (N/(N-K_adj)) invsym(X´X)*P*invsym(X´X)
```
where

```
P = X´diag(e:^2)*X
```
and where

the almost-mathematical notation `diag(e:^2)` specifies the $N \times N$ matrix with the squared elements of **e** along its diagonal

and

X is augmented with a column of 1s if the model includes an intercept

In models with an intercept, **V_robust** can be calculated using the mean-deviated formula

```
V_robust = (N/(N-K_adj)) invsym(S´S)*P*invsym(S´S)
```
where

```
S = X :- mean(X)
```

This formula yields better precision. P could also be calculated using a mean-deviated formula, but we will not bother to do that.

Matrix **P** is commonly known as the outer-product of the gradients.

―――――――――――――――――――――――― Formula sheet 3 ―――――――

Note: This formula sheet uses the same notation as the one in section 11.4.1.

Programming formulas such as

```
V_robust = (N/(N-K_adj)) invsym(S´S)*P*invsym(S´S)
```
and

```
V_opg = (s2)^2 ((N-1)/N) invsym(P)
```

would be daunting if we were starting from scratch. However, we have an up-and-running linear-regression system, and most of the ingredients are already at our fingertips. To implement **Vopg()** and **Vrobust()**, we will need

Mathematical term	Corresponding function returning value
N	N()
K_adj	K_adj()
s2	s2()
invsym(X'X)	XXinv()
invsym(S'S)	SSinv()
e	*not yet implemented*
P	*not yet implemented*

We just need to implement two new subroutines for use by **LinReg::Vopg()** and **LinReg::Vrobust()**, namely,

```
real scalar LinReg::e()
real matrix LinReg::P()
```

e is the vector of residuals, the difference between the observed and predicted outcomes. The new routine for calculating them is

```
real colvector LinReg::e()   return(*y - yhat())
```

I discovered that the existing code already had the calculation for **e** buried in **LinReg::ee()**:

```
real scalar LinReg::ee()
{
        real colvector   e
        if (ee == .z) {
                e  = *y - yhat()
                ee = quadcross(e, e)
        }
        return(ee)
}
```

I changed the boldfaced line to call the new function:

```
e = e()
```

I did not have to change **LinReg::ee()**. But I did change it, because someday I might need to modify how **e** is calculated and forget that the same formula needs updating elsewhere.

The other new calculation we need is P. Calculating it is straightforward using
quadcross():

```
real matrix LinReg::P()
{
        return(quadcross(*X, cons, e():^2, *X, cons))
}
```

We now have the ingredients we need, and it will be easy to write the code for Vopg()
and Vrobust():

```
real matrix LinReg::Vopg()
{
        real scalar    cf
        if (Vopg == .z) {
                cf      = s2()^2 * (N()-1) / N()
                Vopg    = cf * invsym(P())
        }
        return(Vopg)
}

real matrix LinReg::Vrobust()
{
        real scalar    cf
        if (Vrobust == .z) {
                cf = N() / (N()-K_adj())
                if (cons) {
                        Vrobust = cf * (SSinv()*P()*SSinv())
                }
                else    Vrobust = cf * (XXinv()*P()*XXinv())
        }
        return(Vrobust)
}
```

I would say that we are done, but when I later tested the code, I discovered that
Vrobust() did not return a symmetric matrix when it should have. The returned
matrix differed trivially from its symmetric counterpart. The elements on either side
of the diagonal differed from each other by less than 1e–15, and they differed from the
right answer by less than 1e–15, too.

I almost did not notice the problem, and when I did, I thought about ignoring it because
the problem does not arise often, and even when it does arise, it would not matter if you
ignored it. Relative errors of 1e–15 are machine precision, and in general, you should
expect round-off errors larger than this. We nonetheless should discuss the problem,
because there are important numeric accuracy issues that you should know about, and
anyway, it is just plain sloppy to return a nonsymmetric matrix when it should be
symmetric.

I will first modify the `Vrobust()` code, and I will then explain. When a matrix is almost-but-not-quite symmetric *and you know that it is due to round-off error*, you use Mata's `_makesymmetric()` function to make it symmetric. The improved code for `Vrobust()` is

```
real matrix LinReg::Vrobust()
{
        real scalar    cf
        if (Vrobust == .z) {
                cf = N() / (N()-K_adj())
                if (cons) {
                        Vrobust = cf * (SSinv()*P()*SSinv())
                }
                else    Vrobust = cf * (XXinv()*P()*XXinv())
                _makesymmetric(Vrobust)
        }
        return(Vrobust)
}
```

`_makesymmetric(`X`)` makes X symmetric by copying the values from elements below the diagonal to the corresponding elements above the diagonal. Doing this makes the matrix symmetric, but it does not make it any more accurate. Below the diagonal, there was round-off error; above the diagonal, there was round-off error. There is no reason to believe that the error is less on one side or the other. We are replacing the round-off error on one side with the round-off error from the other. Total error is expected to remain the same.

13.4.1 Aside on numerical accuracy: Order of addition

The calculation `SSinv()*P()*SSinv()` does not produce a symmetric matrix when it mathematically should. If we call the resulting matrix `C`, then `C[`i,j`]` is not exactly equal to `C[`j,i`]`, although they are almost equal. The maximum relative absolute difference across all i and j is approximately 1e–15. The difference occurs because the order in which terms are summed when calculating `C[`i,j`]` differs from the order when calculating `C[`j,i`]`.

Addition and subtraction of nonintegers performed by computers are commutative but not associative. That is, $a+b$ always equals $b+a$, numerically speaking, but $(a+b)+c$ is not necessarily equal to $a+(b+c)$. This means that whenever computers add three or more values, the order in which it adds them matters. Think back to section 5.2.1.2 when I had you pretend that you were a base-10 computer with a five-digit accumulator and asked you to sum 5.17 and 0.00014231. You obtained 5.1701. You lost the least significant digits from the second number, and the number of digits you lost depended on the magnitude of the first number. You would have obtained the result 5.1701 even if you had summed the values in the reverse order. Numerical addition is commutative.

If I asked you to sum three numbers, however, the order in which you sum them will affect the result. Sum 5.17, 0.00014231, and 0.00085769. If you sum them in the order specified, and truncate all along the way to five digits, you will obtain 5.1709. Sum the smaller two values first and then add 5.17, however, and you will obtain 5.18, which happens to be the mathematically correct answer.

You can in general minimize the round-off error in any calculated sum by placing the numbers to be summed in ascending order of their absolute values. Programmers rarely do this because real computers work with the equivalent of 16 decimal digits (approximately), and the extra effort seldom results in a meaningful reduction of total round-off error. I say seldom because there are cases in which the extra effort is justified.

13.4.2   Aside on numerical accuracy: Symmetric matrices

Most formulas mathematically guaranteed to produce symmetric matrices involve matrix multiplication, which itself involves summing terms obtained by (scalar) multiplication.

Imagine that $R = AB$ produces a mathematically symmetric result. When the numbers being summed to produce $R_{i,j}$ and $R_{j,i}$ are different but with the mathematical requirement that their respective sums are equal, then R=A*B can produce an asymmetric result because summing different numbers produces different round-off errors. In other cases, the elements being summed on either side of the diagonal are the same, but the order in which they are summed is different, and that difference can lead to different numerical results.

At this point, you may suspect that computers seldom produce numerically symmetric matrices. If you mean seldom in the sense of across all formulas that mathematically produce symmetric matrices, you are right. If you mean seldom in the code that you write, you are probably wrong. Most of the symmetric matrices that statistical programmers code are of the form X'X and X'D*X where D is a diagonal matrix. These two formulas have the property that the elements being summed on the two sides of the diagonal are not only the same but also in the same order. Calculations of X'X and X'D*X always produce numerically symmetric results.

The formula that Vrobust() implements is of the form S'P*S, where all three matrices are symmetric. Mathematically, results will be symmetric. Numerically, they are not guaranteed to be, because P is not a diagonal matrix.

Whenever you code a matrix formula other than X'X or X'D*X intended to produce a symmetric matrix, you need to consider the possibility that the result R will not be symmetric. Here is how you do that. First, code the formula, and then check the result with issymmetric(R) or R==R'. If R is not symmetric, check whether it is nearly symmetric by calculating mreldif(R', R). Depending on how much calculation has gone into producing R, that maximum relative difference will be 1e–15 or as large as 1e–13. It might even be 1e–12. At this point, you are determining whether R is being calculated correctly and the asymmetry could reasonably be caused by round-off error. If it is, use _makesymmetric(R).

I want to explain about 1e–12. Writing a system of equations and having total round-off error of 1e–12 would usually be acceptable. Most consumers of computer calculations (users) think calculated results are more accurate than they are. For instance, in many statistical models other than linear regression, if you succeed in calculating a variance matrix to 12 digits of accuracy compared with truth, you have done an outstanding job. If numeric derivatives are involved, you will be lucky if you have 8 digits of accuracy and, in some cases, even 4 is acceptable.

I suggest 1e–12 for checking symmetry because we are not comparing with truth. We are checking the inaccuracy introduced by one matrix formula. Yes, the ingredient matrices include round-off error, but in checking the calculated result for symmetry, we are treating the ingredients as if they were fully accurate and merely asking how much round-off error the single calculation introduced. A loss of four decimal digits would be concerning, although not overly concerning. If you have round-off error problems, you need to learn about matrix balancing. Balancing reduces the norm of matrices in hopes of improving the accuracy of calculations.

In Mata, it is not necessary to make matrices symmetric that might not be symmetric because of round-off error if you are going to take `invsym()` of them. Do you remember when I showed you that `invsym()` produces wrong answers when given nonsymmetric matrices? It does that so that it can produce correct answers when matrices are nonsymmetric because of round-off error. `invsym()` uses only the elements stored on-and-below the diagonal in making its calculations. If `invsym()` needs to access `x[i,j]` above the diagonal, it accesses `x[j,i]` instead. Thus, the matrix is treated as if it were symmetric, and programmers can ignore the lack-of-symmetry problem in this case. All of Mata's functions that end in `*sym()` have this property.

13.4.3 Finishing the code

We have written new routines `LinReg::e()`, `LinReg::P()`, `LinReg::Vopg()`, and `LinReg::Vrobust()`. All that remains is to satisfy the class's bureaucratic requirements. We must declare the new functions along with their member variables in the class definition. Only `Vopg()` and `Vrobust()` stored their results, so the additions we need to make are

```
class LinReg
{
    public:
            .
            .
        real matrix    Vopg()
        real matrix    Vrobust()
            .
            .

    private:
            .
            .
        real matrix    P()
            .
            .
        real matrix    Vopg
        real matrix    Vrobust
}
```

We also need to add code to set the member variables `Vopg` and `Vrobust` to `.z` in `LinReg::clear()`,

```
void LinReg::clear()
{
            .
            .
        Vopg    = .z
        Vrobust = .z
            .
            .
}
```

We will call `LinReg` with these additions `LinReg` version 2.

13.5   LinReg version-2 code

You can see a complete listing of the `LinReg` version-2 code by typing in Stata

```
. view ~/matabook/linreg2.mata
```

If you have not yet downloaded the files associated with this book, see section 1.4.

13.6   Certifying LinReg version 2

I did not mention it when I presented the code for `LinReg` version 1 earlier in this chapter, but I created a certification script for it. Because `LinReg` version 1 was nothing more than a class translation of the structure-based `lr*()` version-2 code, I also translated its certification script to create `test_linreg1.do`. You can see the file by typing

```
. view ~/matabook/test_linreg1.do
```

I did that before writing `LinReg` version 2 because I did not want to waste my time adding `Vrobust()` and `Vopg()` to code that did not already work.

Then, because I did not want to show you code for `Vrobust()` and `Vopg()` that did not work, I added tests to the bottom of `test_linreg1.do` and saved the result as `test_linreg2.do`. You can see the resulting `test_linreg2.do` by typing

```
. view ~/matabook/test_linreg2.do
```

13.7 Adding LinReg version 2 to the lmatabook.mlib library

Copy the following files to your approved source directory:

```
linreg2.mata
test_linreg2.do
```

Change to your approved source directory. Add the line "do `test_linreg2`" to `test.do`:

```
———————————————————————————————— test.do ———————————
// version number intentionally omitted
do test_n_choose_k
do test_lr2
do test_linreg2
———————————————————————————————— test.do ———————————
```

Run it by typing do `test`.

Add the line "do `linreg2.mata`" to `make_lmatabook.do`:

```
——————————————————————————————— make_lmatabook.do ——————————
// version number intentionally omitted
clear all
do hello.mata
do n_choose_k.mata
do lr2.mata
do linreg2.mata
lmbuild lmatabook.mlib, replace
——————————————————————————————— make_lmatabook.do ——————————
```

Run it by typing do `make_lmatabook`.

Rerun `test.do`.

You are done.

14 Better variable types

14.1 Overview

In Mata, you declare variables in terms of how they are stored, such as

```
real scalar
string vector
transmorphic matrix
```

Code can be more readable if you declared them in terms of how they are used, such as

```
`boolean´ scalar
`Filename´ vector
`Color´ matrix
```

Consider the `EarthDistance::distance()` function discussed in section 12.5. The function was described in terms of how the variables were stored:

```
distance = e.distance(real vector pos1, real vector pos2)
                                 ----                ----
```

The function could have been described by how the variables were used:

```
distance = e.distance(`Position´ pos1 `Position´ pos2)
```

Mata does not itself have features for defining variable types based on use, but Stata has macros that we can use to create the new types.

14.2   Stata's macros

Stata's macros serve as variables in Stata's ado programming language, and many users who do not program use Stata's macros as shorthands to save typing. Let's review how they work.

If you type in Stata

```
. local x 10
```

then `x'—left quote, x, right quote—becomes a synonym for 10. Were you now to type

```
. generate b = a + `x´
```

Stata would see it as

```
. generate b = a + 10
```

Stata sees this because Stata's input processor finds the quoted macros in the input and substitutes their definitions before Stata sees the line. Macros can contain anything and can be used anyplace. For instance, you could define

```
. local G generate
```

and type

```
. `G´ b = a + `x´
```

Stata would see and execute generate b = a + 10.

14.3   Using macros to create new types

Macros defined in Stata can be used in Mata, and just as with Stata, the macros will be found and substituted before Mata sees the input. Consider a Mata function to convert degrees to radians:

```
real scalar radians(real scalar d)
{
        return(d*(pi()/180))
}
```

If you defined in Stata

```
local Degrees   real
local Radians   real
```

you could code **radians()** as

```
`Radians´ scalar radians(`Degrees´ scalar d)
{
        return(d*(pi()/180))
}
```

Because the word **real** is substituted before Mata sees the input, the updated version of the function is indistinguishable from the original from Mata's perspective.

'**Radians**' and '**Degrees**' are called macroed types or derived types.

You can create macroed types that include the organizational type as well:

```
local DegreesS   real scalar
local RadiansS   real scalar
```

Then you could code

```
`RadiansS´ radians(`DegreesS´ d)
{
        return(d*(pi()/180))
}
```

I sometimes do this. I use suffix **S** for scalar, **V** for vector, **RV** for rowvector, **CV** for colvector, and **M** for matrix.

A **.mata** file containing **radians()** with macroed types could read

```
─────────────────────────────────────── add.mata ─────────
version 15
set matastrict on
local DegreesS    real scalar
local RadiansS    real scalar
mata:
`RadiansS´ radians(`DegreesS´ d)
{
        return(d*(pi()/180))
}
end
─────────────────────────────────────── add.mata ─────────
```

The macroed types are defined in the do-file along with the Mata code. They are defined before entering Mata.

14.4   Macroed types you might use

When and how to use macroed types is a personal decision. You use them to increase the readability of code and thereby reduce the chances of errors. You use them so that later, when you return to the code, you will be able to understand it more easily. The flip side is that if the code is already readable using Mata's standard types, there is no reason to complicate it with macroed types.

You might decide that `radians()` was already readable in its original form. Yet, if `radians()` was just one of a set of routines in which some use radians and others use degrees, you might change your mind.

I do not bother with macroed types for naturally numeric variables. For instance, I am fine with declaring subscripts such as `i` and `j` as real scalars when writing matrix functions. I could define a 'Subscript' type, but why bother? I think of `i` and `j` as numbers anyway, and unnecessarily adding new terms does not improve readability.

That said, let me show a matrix example where I would use macroed types for subscripts. In the example below, I am working with a matrix `X` in which the rows are observations and the columns are variables of a dataset. In this case, I might indeed declare subscripts such as 'Obs' and 'Var' just so that I could keep them straight. I would certainly do that if I also had to use a subset of the variables (columns) of `X` specified in vector `v`. `v` might be (1, 32, 41, 89), meaning that the routine is not to use all the columns of `X`; it is to use just the columns 1, 32, 41, and 89. I could code

```
local Obs real
local Var real

mata:

real scalar myprogram(real matrix X, `Var´ vector v)
{
        `Obs´ scalar      i
        `Var´ scalar      j
        real scalar       cj

        for (cj=1; cj<=length(v); cj++) {
                j = v[cj]
                for (i=1; i<=rows(X); i++) {
                        ...  X[i,j] ...
                }
        }
        ...
        return(...)
}

end
```

What is important in the above code is the declaration of v. Had it been declared as a real vector, it would appear to be just a vector. Well, it is a vector, but it is a vector of subscripts, and row subscripts at that. I jumped all the way to what the row subscripts mean and called the type Var. Perhaps I went too far. An equally good solution would have been

```
local RowSub    real
local ColSub    real
```

There is no right answer, but in this case there might be a wrong answer, which would have been to declare all the variables as real.

I told you about the capital letter suffixes S, V, RV, CV, and M that I sometimes use. Had I used them, the program would have read as follows:

```
local ObsS real scalar
local VarS real scalar
local VarV real vector

mata:
real scalar myprogram(real matrix X, `VarV´ v)
{
        `ObsS´ scalar    i
        `VarS´ scalar    j
        real scalar      cj

        for (cj=1; cj<=length(v); cj++) {
                j = v[cj]
                for (i=1; i<=rows(X); i++) {
                        ...  X[i,j] ...
                }
        }
        ...
        return(...)
}
end
```

When I use macroed types, I usually define shorter names for Mata's standard types. I define 'RS' for real scalars, 'RV' for real vectors, and so on:

```
local RS   real scalar
local RV   real vector
local RM   real matrix
```

If I need strings, I do the same for the string types that I need:

```
local SS   string scalar
local SV   string vector
```

Just using 'RS', 'RV', and 'SS' can make code more readable.

The code for `myprogram()` would now read

```
// ------------------------------------- standard types
local RS    real scalar
local RV    real vector
local RM    real matrix

// ------------------------------------- derived types
local ObsS `RS´
local VarS `RS´
local VarV `RV´

`RS´ myprogram(`RM´ X, `VarV´ v)
{
        `ObsS´     i
        `VarS´     j
        `RS´       cj

        for (cj=1; cj<=length(v); cj++) {
                j = v[cj]
                for (i=1; i<=rows(X); i++) {
                        ... X[i,j] ...
                }
        }
        ...
        return(...)
}
```

I have no hard and fast rules. Sometimes I use Mata's types directly. Sometimes I use macroed types. When I do use macroed types, sometimes I use them one way and sometimes another. My goal is to increase the readability of the code.

I do have a few favorite macroed types that I can recommend to you, however.

14.4.1 The boolean type

The macroed type I use most often is 'boolean', and I also define the values True and False to go along with it.

```
local boolean    real
local True       1
local False      0
```

To see why I like this type, I ask you what the following code does.

```
real scalar fullrank(real matrix X)
{
        real matrix        Xinv

        Xinv = invsym(X)
        for (i=1; i<=rows(X); i++) {
                if (X[i,i]==0) return(0)
        }
        return(1)
}
```

Now look at the same code after I substitute 'boolean', 'False', and 'True' for real, 0, and 1:

```
`boolean´ scalar fullrank(real matrix X)
{
        real matrix        Xinv

        Xinv = invsym(X)
        for (i=1; i<=rows(X); i++) {
                        if (X[i,i]==0) return(`False´)
        }
        return(`True´)
}
```

You now see right away that the function returns 'True' or 'False'. It returns whether a symmetric matrix is of full rank.

14.4.2   The Code type

Another macroed type I use frequently is

```
local Code        real
```

I use 'Code' for variables containing numeric codes, as in

```
real scalar foo(real colvector y, real matrix X,
                `Code´ scalar method)
{
        .
        .
        if (method==1) {
                ...
        }
        else if (method==2) {
                ...
        }
        else if (method==3) {
                ...
        }
        else    _error(3300, "invalid method")
        .
        .
}
```

When I use 'Code', I usually define macros for the code's values, too:

```
local Code              real
local MaxDescent          1
local Newton              2
local ConjugateGradient   3
```

Then my program can read

```
real scalar foo(real colvector y, real matrix X,
                `Code´ scalar method)
{
        .
        .
        if (method==`MaxDescent´) {
                ...
        }
        else if (method==`Newton´) {
                ...
        }
        else if (method==`ConjugateGradient´) {
                ...
        }
        else    _error(3300, "invalid method")
        .
        .
}
```

I sometimes write programs that have more than one set of codes, and in that case, I distinguish between them by giving them different types:

```
local MethodCode              real
local M_MaxDescent            1
local M_Newton                2
local M_ConjugateGradient     3

local TypeCode                real
local T_expr                  1
local T_fcn                   2
```

Code to me means a cardinal code—codes that have no ordering. The codes above are cardinal. `TypeCode` is coded 1 and 2, but it could just as well be coded 2 and 1. Codes where order matters are called ordinal, and I use the suffixes `Ordinal` or `Ocode` for them. In what follows, I am writing code to maximize a user-specified function, and I want to record whether the user also supplied the functions to calculate the first and second derivatives:

```
local Fdef_Ordinal    real
local F0              0    // F() defined
local F1              1    // F(), F´() defined
local F2              2    // F(), F´(), F´´() defined
```

If variable `f` is an 'Fdef_Ordinal', it would be reasonable to ask whether f>'F0'.

14.4.3   Filehandle

File handles are the codes that Mata's file I/O functions use to track files. They are stored as real scalars, but I define

```
local Filehandle      real
```

so that I can write code like this:

```
string scalar get_first_line(string scalar filename)
{
        `Filehandle´ scalar    fh
        string matrix          line

        fh = fopen(filename, "r")
        line = fget(fh)
        fclose(fh)
        return(line)
}
```

14.4.4   Idiosyncratic types, such as Filenames

Idiosyncratic types means peculiar to the particular program I am writing. Variables containing filenames is a good example. To me, a filename is just a string scalar if I have just one or two of them. If I am writing routines to manipulate or manage hundreds of filenames, however, I would define

```
local Filename      string
```

I would define 'Filename' so that I could quickly distinguish the 'Filename's in the code from the other string variables.

14.4.5   Macroed types for structures

I invariably assign macroed types to structures. Consider the following structure:

```
struct Earth_Position
{
        real scalar        latitude
        real scalar        longitude
        `boolean´ scalar   degrees_not_radians
}
```

The function **path()** uses **struct Earth_Position** variables:

```
struct Earth_Position vector path(
        struct Earth_Position scalar here,
        struct Earth_Position scalar there,
        struct Earth_Position vector obstacles)
{
        ...
}
```

The function declaration is difficult to read because there is too much ink dedicated to the structures. Beyond that, I object to the presence of the word **struct** because it magnifies what to me is a detail. To me, an **Earth_Position** is a type just as much as **real**, **string**, and 'boolean' are types. **Earth_Position** is not a lesser type for having been implemented as a structure. The situation can be much improved if we define

```
local Position      struct Earth_Position
local PositionS     `Position´ scalar
local PositionV     `Position´ vector
```

The program then would read

```
`PositionV´ path(`PositionS´  here,
                `PositionS´ there, ` PositionV´ obstacles)
{
        . . .
}
```

It is now obvious what `path()` does. It returns a path from `here` to `there` avoiding `obstacles`.

14.4.6   Macroed types for classes

I object to including the word `class` in declarations just as much as I object to including `struct`. It is a detail of implementation, and one that I am not likely to forget, so it does not need emphasis. If '`Position`' was a class instead of a structure, I still would have coded

```
`PositionV´ path(`PositionS´  here,
                `PositionS´ there, `PositionV´ obstacles)
{
        . . .
}
```

The only difference would be the macroed type definitions:

```
local Position        class Earth_Position
local PositionS       `Position´ scalar
local PositionV       `Position´ vector
```

Many classes are inherently scalars, such as `LinReg` in the previous chapter. As a user of `LinReg`, I would define '`LinReg`' to be a class `LinReg` scalar:

```
local LinReg          class LinReg scalar
```

Then I would use it like this:

```
... my_linear_regression(...)
{
        `LinReg´    r
          .
          .
          .
}
```

14.4.7   Macroed types to avoid name conflicts

With the exception of Mata objects declared in ado-files, the names we give to Mata functions, classes, and structures are public, and that means we must choose names that others have not chosen. We could not name a function `distance()`, for instance.

Someone else is bound to have used that name, and even if no one has yet, StataCorp might someday create a `distance()` function.

In section 12.5, we discussed the use of function-only classes as a way of giving functions more meaningful names while still avoiding name conflicts. We created a `distance()` function in class `EarthDistance`, a slightly inelegant name that we chose to avoid name conflicts. Users can declare `EarthDistance` variables and so access the `distance()` function:

```
... some_users_function(...)
{
        class EarthDistance scalar    e
        .
        .
        ... e.distance(...)
        .
        .
}
```

Problem solved. If some argued that `EarthDistance` is likely to result in name conflicts, we could choose an even longer, more ungainly name, such as

Goulds_Earth_Distance_Functions

and users would not be much inconvenienced. The above code would now read

```
... some_users_function(...)
{
        class Goulds_Earth_Distance_Functions scalar e
        .
        .
        ... e.distance(...)
        .
        .
}
```

If users found themselves wanting to use Goulds_Earth_Distance_Functions in lots of routines, they could create a macroed type for it to escape typing the long declaration:

```
local Edist        class Goulds_Earth_Distance_Functions
```

Then their code could read

```
... some_users_function(...)
{
        `Edist´ scalar    e
        .
        .
        ... e.distance(...)
        .
        .
}
```

We at StataCorp use the ungainly name method for classes and structures that we do not wish to make public but which we do need to use as subroutines in code that is made public. We might create a class named **GPS_GO_Graph_output**. That might be the class's name, but we would not use it in code. We would create a macroed type:

```
local Gout        class GPS GO Graph_output
```

We would use 'Gout' in our code.

In fact, we use the macroed-name approach even with code that is intended to be made public. In the previous chapter, we developed **LinReg**. We would never have done that at StataCorp even if we believed that the final name would be **LinReg**. We would have created **LinReg_development_project** or some other long name. Despite that, the .mata files would have looked nearly the same as the ones I showed you. File **linreg2.mata** would have read

────────────────────────── alternative `linreg2.mata` ──────────

```
version 15
set matastrict on
local LinReg    LinReg_development_project
mata:
class `LinReg´ {
        .
        .
}

void `LinReg´::clear
{
        .
        .
}

.
.
```

────────────────────────── alternative `linreg2.mata` ──────────

Even the certification do-file would have used macroed types:

```
                                    ─── alternative test_linreg2.do ───
// version number intentionally omitted
clear all
run linreg2.mata
local LinReg   LinReg_development_project
    .
    .
mata:
    .
    .
r = `LinReg´()
r.setup(y, X, 1)
    .
    .
                                    ─── alternative test_linreg2.do ───
```

Later, we would have had a meeting where we would decide on the final name for the class. If **LinReg** was to be the final name, the macro definitions in **linreg2.mata** and **test_linreg2.do** would change to be

```
local LinReg   LinReg
```

If **LinearRegression** was to be the final name, the line would change to

```
local LinReg   LinearRegression
```

Names are too important to be chosen as you write the first lines of code. You need to think about them. You should write code so that you can easily change names right up until the day you release. After that, you are stuck with them.

15 Programming constants

15.1 Problem and solution

Constants are numbers that appear in programs that specify physical values, tolerances, limits, and the like. Constants cause three problems.

First, they can make code difficult to read. Consider the following calculation:

```
ly = 2.9998e+8*60*60*24*365.25
```

Is it obvious to you that `ly` is a light year?

Second, constants are sometimes not constant. The limit of 100 that makes sense today may need to be larger tomorrow.

Third, constants in code are difficult to track down when you need to change them. Consider the 100 limit that you want to increase. You will discover that the 100s you need to change look just like every other 100 that appears in the code.

The solution to all of these problems is to store constants in Stata macros and reference the macros. Mata can understand the line

```
ly = `speed_of_light_meters_per_sec' * `secs_per_year'
```

if you have previously defined

```
local speed_of_light_meters_per_sec    (2.9998e+8)
local secs_per_year                    (60*60*24*365.25)
```

If you have a limit of 100, you can define

```
local max_values        (100)
```

and refer to 'max_values' in the code.

If you have a blocking factor of 100, you can define

```
local blocking_factor    (100)
```

and refer to 'blocking_factor' in the code. The fact that 'blocking_factor' and 'max_values' happen to both be 100 is irrelevant. If you need to increase one but not the other, change the macro definition and recompile.

15.2   How to define constants

Define constants using

```
local name (definition)
```

where *definition* is a numeric or string literal or expression. Examples include

```
local speed_of_light_meters_per_sec   (2.9998e+8)
local secs_per_year        (60*60*24*365.25)
local header               ("   N        mean        S.D.\n")
local divider              (31*"-")
```

The parentheses can be omitted in the case of numeric literals, but do not do that. If you become accustomed to seeing definitions like

```
local speed_of_light_meters_per_sec   2.9998e+8
```

you might omit the parentheses in the other cases when the consequences would be dire. Consider secs_per_year defined without the parentheses and having subsequent code that reads

```
tsquared = `secs_per_year'^2
```

The expanded line would then read

```
tsquared = 60*60*60*24*365.25^2
```

Only the 365.25 would be squared! When the macro is defined with the parentheses, the result will be as desired:

```
tsquared = (60*60*60*24*365.25)^2
```

15.3   How to use constants

You use the macros in code in the standard way—open single quote, macro name, close single quote—such as 'secs_per_year'. You can code

```
ly = `speed_of_light_meters_per_sec' + `secs_per_year'
tsquared = secs_per_year'^2
printf(divider)
```

15.4   Where to place constant definitions

You place constant definitions at the top of .mata files, either before or after any macroed types you create:

```
────────────────────────────────────── filename.mata ──────────
version 15
set matastrict on
local RS        real scalar
local RV        real vector
local boolean   real
local True      1
local False     0
local speed_of_light    (2.9998e+8)
local secs_per_year     (60*60*24*365.25)
local max_values        (100)
local blocking_factor   (100)
local header            ("      N          mean          S.D.\n")
local divider           (31*"-")
 mata:
 .
 .
 .
 end
────────────────────────────────── filename.mata ──────────
```

You place them later in ado-files, but still just above the Mata code:

```
——————————————————————————————— filename.ado ——————————
*! version 1.0.0

program ...
        version 15
            .
            .
end

    .
    .
version 15
set matastrict on

local RS         real scalar
local RV         real vector
local boolean    real
local True       1
local False      0
local speed_of_light    (2.9998e+8)
local secs_per_year     (60*60*24*365.25)
local max_values        (100)
local blocking_factor   (100)
local header            ("      N          mean          S.D.\n")
local divider           (31*"-")

  mata:
    .
    .
  end
——————————————————————————————— filename.ado ——————————
```

16 Mata's associative arrays

16.1 Introduction

Array is computer jargon for vectors, matrices, and super matrices.

A one-dimensional array is a vector with elements $A[i]$.

A two-dimensional array is a matrix with elements $A[i, j]$.

A three-dimensional array is a super matrix with elements $A[i, j, k]$. And so on.

An associative array is an array in which the indices are not necessarily integers. Most commonly, the indices are strings, so you can have $A["bill"]$, $A["bill", "bob"]$, or $A["bill", "bob", "mary"]$. The indices do not have to be strings. Another associative array might have elements $B[1.25]$, $B[1.25, 3]$, or $B[1.25, 3, -5.2]$.

Mata has a class named AssociativeArray that implements these features. I want to tell you about it because it is such a useful programming tool. I also want to emphasize that it was written in Mata. It is not an internal feature coded in C. It is a feature that we added to Mata in the same way as I describe in this book. We at StataCorp wrote code in .mata files. We wrote certification do-files. We compiled the code and certified it. We finally placed it in a Mata library so that others could use it.

We wrote the code before Mata had classes. We implemented associative arrays using structures and functions with related names. The function names all began with the prefix asarray, just as in chapter 11, the lr*() functions all started with the letters lr. Later, we created a class. We did not reimplement the code. We instead wrote a class that used the existing asarray*() functions.

I will tell you where you can learn more about the implementation, but first, I want to tell you how to use AssociativeArray.

16.2   Using class **AssociativeArray**

Associative arrays are created when you code

```
class AssociativeArray scalar   A
```

or, if working interactively,

```
A = AssociativeArray()
```

Either way, A is now an associative array. It is created as a one-dimensional array with string keys (indices) by default. You can change that. If you wanted a two-dimensional array with real-numbered keys, you would use **A.reinit()** to change the style. You would code

```
A.reinit("real", 2)
```

The **"real"** you specify is not the type of A's elements. It is the type of A's keys. You can store whatever types you want in A's elements, and you can store different types in different elements. Not even Mata's transmorphic matrices allow that. Transmorphic matrices can be of any type and they can change their type as the code executes, but at any instant, the elements stored in the matrix are all of the same type. A is different. One element could be a real scalar, another a real matrix, another a string scalar, and so on. An element could even be a class **LinReg** scalar or even another **AssociativeArray**!

The statement **A.reinit("real", 2)** makes A two-dimensional with type real indices. A is like a matrix in that we can store elements $A[1, 2]$ and $A[2, 1]$, which we can do by using **A.put()**:

```
A.put((1,2),   5)
A.put((2,1),  "this")
```

Note that the indices are specified inside an extra set of parentheses. That is, we typed

```
A.put((1,2), 5)
```

and not

```
A.put(1,2,  5)
```

A.put() is a function that takes two arguments, the index and the value to be stored. **A.put()** expects the first argument to be a 1×2 vector because we reinitialized A to be two-dimensional. In the jargon of associative arrays, $(1, 2)$ and $(2, 1)$ are called keys.

Although associative arrays allow elements to be of different types, such usage is rare. Let's start over and store all numeric values.

```
: A = AssociativeArray()

: A.reinit("real", 2)
: A.put((1,2),   5)
: A.put((2,1),   2)
```

Oops, I made a mistake. I meant to set the $(2, 1)$ element to -2 instead of 2. I can fix that:

```
: A.put((2,1),  -2)
```

Stored values can be retrieved using the **A.get()** function. If we typed

```
: x = A.get((1,2))
: y = A.get((2,1))
```

then x would equal 5 and y would equal -2. It may seem as if A is a regular matrix, but it is not. It is not a regular matrix because $A[1, 1]$ and $A[2, 2]$ simply do not exist. If we fetched the value of $A[1, 1]$ by typing

```
: z = A.get((1,1))
```

z would equal J(0,0,.), which is to say, a real 0×0 matrix. This is **AssociativeArray**'s way of saying that $A[1, 1]$ is undefined. Returning J(0,0,.) is the default, just as the keys being one-dimensional strings was a default. We can change what is returned using the **A.notfound()** function. We could make the not-found value be 0:

```
: A.notfound(0)
```

Now **A.get((1,1))** will return 0. We are on our way to creating sparse matrices! Sparse matrices are matrices in which most of the values are 0. Matrix A has two nonzero values. I do not know whether the matrix is 2×2 or 1000×1000 because $A[i, j]$ is 0 for all (i, j) not equal to $(1, 2)$ or $(2, 1)$. If we wanted to treat A as a sparse matrix, we would need to keep track of the overall dimension separately.

A's sparseness is not just superficial. If we defined

```
: A.put((1000,1000),  6)
```

then the A "matrix" would be at least 1000×1000. Yet, because it is an associative array, only three elements are stored! A regular Mata 1000×1000 matrix would consume 7.8 megabytes of memory.

Storing sparse matrices is just one use of associative arrays. Most applications of associative arrays use one-dimensional string keys, which is why that is the default. The classic example is a dictionary.

```
: Dictionary = AssociativeArray()
```

The words in the dictionary are the keys, and their definitions are the elements. Here is an example:

```
: Dictionary.get("aback")
  [1, 1] = (archaic) toward or situation to the rear.
  [2, 1] = (sailing) with the sail pressed backward against
                     the mast by the head.
```

When I created this dictionary, I stored each definition as a string column vector. I put the first definition in the first row, the second definition in the second (if there was one), and so on. Here is exactly what I typed to enter the two definitions for *aback*:

```
: Dictionary.put("aback",
>              (
>                      "(archaic) toward or situation to the rear."
>              \
>                      "(sailing) with the sail pressed backward against
>                       the mast by the head."
>              ))
```

I have a list of 25,592 words on my computer used by my spelling checker. The great feature of associative arrays is that I could enter definitions for all of them, and still `Dictionary.get()` could find the definition for any of them almost as quickly as it found *aback* when it was the only word stored. Nor does performance depend on entering words in a particular order, such as alphabetically.

16.3   Finding out more about AssociativeArray

You can learn more by typing `help mata AssociativeArray`. The source code is available, too. Scroll to the bottom of the file, and you will see

Source code

```
associativearray.mata
asarray.mata
```

You can click on the filenames listed to see the source.

17 Programming example: Sparse matrices

17.1 Introduction

I am about to show you the entire development process from start to finish, and I will show it to you not in the sterilized form that would usually make it into books. I will show it to you with all the bad ideas, mistakes, and errors that usually occur during development. I do this not to tax your patience, but because bad ideas, mistakes, and errors are inevitable. I want to show you how to solve them or, in some cases, just work your way around them. I want to show you how to write code in the certain knowledge that problems will arise and code that is amenable to being fixed.

We are going to consider a project so large that it needs a formal design. This project is also simple enough that it might deceive you into thinking that a formal design is unnecessary. In my experience, most projects start with deceptively simple ideas, and it is easy to confuse ideas with designs.

It will take us two chapters to work our way though the development process. Sometimes it will seem as if we are slogging through a mire of details. Programming is about details. If details do not interest you, you need to find another line of work. Nonetheless, the particular details of the programming project I have selected may not interest you. If so, just read through. Pay attention to the gestalt of what I am doing. Successful programming is about not losing sight of the goal. You will learn that successful programming is also about writing programs in a style that hides details. If details change, programs change, but the overall approach survives the changes.

In this chapter, I will tell you about the idea, then we will turn it into a design, and finally, we will write initial code.

17.2   The idea

I mentioned in the previous chapter that class `AssociativeArray` could be used to implement sparse matrices. A sparse matrix is a matrix in which most of the elements are 0. Sparse matrices can have hundreds or thousands of rows and columns, meaning they have millions of elements, but only a fraction of them are nonzero. If you could store the nonzero elements only, the memory savings would be considerable. Associative arrays can do that. The challenge is to package the solution so that the matrices are just as convenient to use as regular matrices.

I outlined in the previous chapter how class `AssociativeArray` could be used to implement the storage of sparse matrices:

If `A` is a `class AssociativeArray scalar`, then begin by making it a matrix with real subscripts (keys) and setting the not-found value to 0:

```
A.reinit("real", 2)
A.notfound(0)
```

Now think of `A` as a sparse matrix $A[i, j]$. You can insert values into the matrix by coding as follows:

Code	Meaning
`A.put((1,2), 5)`	$A[1, 2] = 5$
`A.put((2,1), -2)`	$A[2, 1] = -2$
`A.put((1000,1000), 6)`	$A[1000, 1000] = 6$

You can retrieve the stored values:

Code	Meaning
`w = A.get((1,2))`	$w = A[1, 2]$
`x = A.get((2,1))`	$x = A[2, 1]$
`y = A.get((1000, 1000))`	$y = A[1000, 1000]$

If you retrieve undefined elements such as $A[200, 300]$, returned will be 0:

Code	Meaning
z = A.get((200,300))	$z = A[200, 300]$

z will be 0.

The above is the idea.

From this idea, we are going to design a class that allows creation of sparse matrices, storing and retrieving values, and more, including matrix multiplication and addition.

17.3   Design

The code you write is called a system when there is more than one related function that users will call. Having 2 functions does not really qualify as a system except in a formal sense, but if there are 10 or more functions, it certainly qualifies. The class we will be writing will have a lot more than 10 functions. Issues arise in writing systems beyond the choice of algorithms. The primary issue is how the functions fit together.

We have already programmed one system, namely, linear regression. And we programmed it twice, first using structures and then again using classes. The design was the same in both implementations, and that design could be summarized in one short phrase: self-threading code. Organization of the code—how the functions fit together— was determined by that choice.

Self-threading code is not the solution to all programming problems.

17.3.1   Producing a design from an idea

The idea outlined above is a design of sorts. There are unspecified details, but even so, you could start writing code based on the idea and refine the design as you go. That approach works well enough when writing small systems. We have all had the experience of writing a routine and realizing that there is an issue that we had not considered. We deleted and rewrote a half-dozen lines. When writing larger systems, that same experience can involve rewriting much more than a half-dozen lines of code. It can make irrelevant everything you have written. The purpose of a design is to identify and resolve those issues before writing the code.

Think of design as being a dry run. Rather than writing completed code, you write incomplete sketches of it. Writing sketches forces you to think about the issues, or at least the big issues. Sketches will not prevent you from going down blind alleys, but they will make it less costly when you do. Modifying or even discarding sketches is a minor cost compared with modifying or discarding code. And when you have completed the process, the sketches make it easier to produce the final code.

You start by pretending that the system already exists and imagine yourself using it. The systems we are discussing in this book are systems written in Mata for use by other Mata programmers, so you imagine using the as-yet-unwritten Mata functions. When you are satisfied with them as a user, you ask yourself, could I write code to implement them? You make sketches of code to convince yourself that you could. These sketches will fix ideas. As you do this, you will eventually wander down a mistaken path. It cannot be avoided. You will have ignored something important or assumed something that is not true. There is no predicting when you will discover the problem, but you will, and you will backtrack.

Backtracking is easier because you have not invested as much in the details. You will not be so attached to what you have done that you forge ahead anyway. This results in arriving at a better outcome with a better design in hand.

We already have a vague design for a sparse-matrix system. Class SpMat will contain a class AssociativeArray, and SpMat will provide functions to manipulate that array. SpMat will set the associative array to use real 1×2 vectors as keys. Those vectors will be the row-and-column indices. SpMat will store the nonzero values in the array and set the array to return 0 when an element is undefined. Meantime, the user of SpMat will have no idea that AssociativeArray is involved in the solution. The user will see a clean, straightforward implementation of sparse matrices.

Let's imagine SpMat in use. If we used it to create a $1{,}000 \times 1{,}000$ matrix named S, I imagine that we start like this:

```
class SpMat scalar    S
S.setup(1000, 1000)
```

At this point, each element of S will "contain" 0, which I put in quotes because S will really contain nothing. In our code, we will have told AssociativeArray to return 0 when an element is undefined.

One issue, a minor one, is that the class will need to track the total number of rows and columns in the matrix. S.setup() will set the number of rows and columns. Already, we have the start of a partial sketch that defines the class SpMat:

```
class SpMat
{
        public:
            void        setup()
            .
            .
        private:
            class AssociativeArray scalar    AA
            real scalar                      rows, cols
            .
            .

}
```

The vertical ellipses in the sketch would appear in your notes just as they appear in mine. You do not yet know what other functions or variables the class will contain.

We can make a sketch of **S.setup()**:

```
void SpMat::setup(rows, cols)
{
        AA reinit("real", 2)
        AA.notfound(0)

        this.rows = rows
        this.cols = cols
}
```

The above is a sketch, not completed code. I did not bother to specify the variable types of the arguments because it is obvious to me that they are real scalars. We are writing the sketch for ourselves, not the computer.

I write more sketches at the beginning of a design than I will write later because, once I know that I am on track to thinking clearly, sketches become less necessary.

To set $S[1,2] = 5$, we might now imagine that the user of the system will type

```
S.put(1,2, 5)
```

To obtain the value of $S[1,2]$, we imagine the user will type

```
value = S.get(1,2)
```

So far, so good. Let's sketch **S.put()** and **S.get()**:

```
void SpMat::put(i, j, v)
{
        AA.put((i,j), v)
}
void SpMat::get(i, j)
{
        return(AA.get((i,j)))
}
```

I sketched the above and then realized that the code for **S.put()** was inadequate. The user of **S.put()** might call it with a value of 0, and we are not to store the 0s in the associative array. I fixed my sketch of **SpMat::put()**, and I left **SpMat::get()** unchanged below it:

```
void SpMat::put(i, j, v)
{
        if (v) AA.put((i,j), v)
}
void SpMat::get(i, j)
{
        return(AA.get((i,j)))
}
```

The improved sketch stores v only if v is not 0. I did that, and then I realized that
I had solved one problem only to create another. Imagine that the user executes
S.put(1,5, 2) and later executes S.put(1,5, 0). As the code is now written, typing
S.put(1,5, 0) will do nothing. To solve the problem, I thought about changing the
sketch of S.put() to be

```
void SpMat::put(i, j, v)
{
        if (v) AA.put((i,j), v)
        else   AA.remove((i,j))
}
```

AA.remove() is the AssociativeArray function to delete an element. I checked the
documentation on associative arrays and verified that it is allowed to remove an element
even if the element does not exist.

Anyway, I thought about doing the above and then I rejected it. SpMat::put() will
be used to define sparse matrices, and I can easily envision the put() function being
called with lots of 0 values. That would work, but how much time would the function
waste removing elements that did not exist? I could have done timings to find out, but
I did not. I decided there would be another function, replace(), for changing already
stored values. I changed my sketch yet again, this time to read as follows:

```
void SpMat::put(i, j, v)
{
        if (v) AA.put((i,j), v)
}
void SpMat::replace(i, j, v)
{
        if (v) AA.put((i,j), v)
        else   AA.remove((i,j))
}
```

Perhaps you would have resolved the issue differently. That is unimportant. You think
about the issues and resolve them.

The system now includes the following functions:

```
S.init(rows, cols)
S.put(i, j, v)
v = S.get(i, j)
S.replace(i, j, v)
```

As I imagined myself using this system, I thought it would be convenient to be able to define an entire row or column of the matrix from an existing vector. I momentarily considered modifying `S.put()` so that the user could code

    S.put(i, ., *rowvector*)

and

    S.put(., j, *colvector*)

I did not sketch any code. I decided against using `put()` in this way because I wanted to keep `put()` fast, and that meant keeping it simple. I decided instead to create yet another function named `S.Put()` to accompany `S.put()`. `S.Put()` would have the following syntaxes:

Function	Result
`S.Put(i, j,` *scalar*`)`	same as `S.put()` (*)
`S.Put(i, .,` *rowvector*`)`	put row vector
`S.Put(., j,` *colvector*`)`	put column vector
`S.Put(., .,` *matrix*`)`	put entire matrix (*)

(*) write code only if convenient

When I first wrote this table for myself, it included only the middle two syntaxes; I added the other two later along with a note to write the code for them only if convenient. The extra syntaxes are not necessary because putting a scalar can be done with `S.put()`, and as for putting an entire matrix, if users are willing to create a regular matrix containing the sparse values, why are they using `SpMat` in the first place? Nonetheless, I thought, having `S.Put(`*i, j, scalar*`)` be consistent with `S.put()` would be a nice touch and having `S.Put(., .,` *matrix*`)` would be useful for testing the code later.

I also realized that I would want capital-letter variants of `S.replace()` and `S.get()`, so I added to my design:

Function	Result
`S.Replace(i, j,` *scalar*`)`	same as `S.replace()` (*)
`S.Replace(i, .,` *rowvector*`)`	replace row
`S.Replace(., j,` *colvector*`)`	replace column
`S.Replace(., .,` *matrix*`)`	replace matrix (*)
scalar `= S.Get(i, j)`	same as `S.get()` (*)
rowvector `= S.Get(i, .)`	get row
colvector `= S.Get(., j)`	get column
matrix `= S.Get(., .)`	get matrix (*)

(*) write code only if convenient

I finally realized that users would want to get, put, and replace submatrices, too. Does that mean yet another set of functions? No, I decided, I will add yet more features to `S.Put()`, `S.Replace()`, and `S.Get()`. I reasoned that sometimes users would want to work with individual elements. The lowercase functions `S.put()`, `S.replace()`, and `S.get()` would allow that, and they would run as quickly as feasible. Other times, users would want to work with a group of elements, and in those cases, I could afford the bit of computer time to diagnose what needed to be done. Thus, the syntax for `S.Put()` would be

Function	Result
`S.Put( i, j, `*scalar*`)`	same as `S.put()`
`S.Put( i, ., `*rowvector*`)`	put row vector
`S.Put( ., j, `*colvector*`)`	put column vector
`S.Put( ., ., `*matrix*`)`	put entire matrix
`S.Put(i1,j1, i2,j2, `*submat*`)`	put submatrix ← *new*

The last line is new, and in this case, `S.Put()` will have five arguments instead of the usual three. I also decided at this point that implementing the first and fourth syntaxes was no longer optional.

My final specifications for `S.Replace()` and `S.Get()` were similarly modified:

Function	Result
`S.Replace( i, j, `*scalar*`)`	same as `S.replace()`
`S.Replace( i, ., `*rowvector*`)`	replace row vector
`S.Replace( ., j, `*colvector*`)`	replace column vector
`S.Replace( ., ., `*matrix*`)`	replace entire matrix
`S.Replace(i1,j1, i2,j2, `*submat*`)`	replace submatrix

Function	Result
scalar `= S.Get( i, j)`	same as `S.get()`
rowvector `= S.Get( i, .)`	get row
colvector `= S.Get( ., j)`	get column
matrix `= S.Get( ., .)`	get matrix
submat `= S.Get(i1,i2, j1,j2)`	get submatrix

I did not bother to sketch the code for `S.Put()`, `S.Replace()`, or `S.Get()` because I know I could write it. Had I been uncertain, I would have sketched the code for them.

There will be other routines we will want, but they can be easily written as we write the code. For instance, we will want routines to return the rows and columns of a matrix, `S.rows()` and `S.cols()`. The code would be

```
———————————————————————— Sketches of SpMat::rows() and cols() ————————
real scalar SpMat::rows()   return(rows)
real scalar SpMat::cols()   return(cols)
———————————————————————— Sketches of SpMat::rows() and cols() ————————
```

17.3.2   The design goes bad

I then began thinking about matrix multiplication, and as I did, I went down a mistaken path.

A user of `SpMat` will want to multiply sparse matrices, sparse and regular matrices, and regular and sparse matrices. In each case, they will want to be able to store the result as a regular or a sparse matrix. I made a table:

Need to be provided	Shorthand notation
regular = sparse * sparse	R = S1*S2
regular = sparse * regular	R = S1*R2
regular = regular * sparse	R = R1*S2
sparse = sparse * sparse	S = S1*S2
sparse = sparse * regular	S = S1*R2
sparse = regular * sparse	S = R1*S2
sparse = regular * regular	S = R1*R2 (*)

(*) Perhaps unnecessary

I decided that each capability would be provided by a separate function. I then realized that users sometimes need to multiply transposed matrices. Users will sometimes want to calculate, for instance,

```
R = S1´ * S2
```

One solution would be for them to code

```
S2t = S1.transpose()
R   = S1.multiplysparse(S2t)
```

I rejected that because I expect that most sparse matrices will be large. The main reason we are implementing `SpMat`, after all, is to deal with large matrices efficiently. One implication is that users should not need to create additional matrices, even temporarily. If transposition is needed, the code for multiplication can interchange row and column

subscripts. Handling transposition in the code brings its own problems, however. The
single case

```
    R = S1 * S2
```

becomes four cases:

```
    R = S1  * S2        (multiply S1 by S2)
    R = S1´ * S2        (multiply S1 transpose by S2)
    R = S1  * S2´       (multiply S1 by S2 transpose)
    R = S1´ * S2´       (multiply S1 transpose by S2 transpose)
```

This one-case-becomes-four situation applies equally to the other six cases, too, which
means 28 functions to write.

And then I thought about matrix addition and realized that the same issues would
arise. Implementation using separate functions would mean that I would have to write
56 functions in all! At this point, I started looking for a different solution.

Before I explain that different solution, I want to emphasize the benefit of designing
the system before writing it. Let's imagine what would have happened had I skipped
the design and proceeded directly to writing final code. I am a highly experienced
programmer, and I am perfectly capable of writing SpMat from just the original idea.
And I can make it work.

So let us imagine that SpMat is written up to the point of multiplication, and I am
just starting to write the multiplication routines. Oh, I would say to myself, I need to
multiply a sparse by a sparse. And I would write that routine. I do not know which
I would have written first, R = S1*S2 or S = S1*S2, but I would have written one,
realized that I needed the other, block copied the first routine, and quickly modified it
to return the other kind of matrix.

Then I would realize that I needed S1*R2. I would have block copied the two written
routines and modified them to produce the desired results, R = S1*R2 and S = S1*R2.
Oops, I need R1*S2, too. Well, this is getting tedious, but I can block copy and modify
the two S1*R2 routines. And so I would have continued until I had all 28 functions
written.

Then I would discover I had a bug, and, mumbling in irritation, I would have fixed each
of the 28 functions. That done, I would have block copied the 28 multiplication routines
to create another 28 functions, which I would modify to handle addition. I would have
realized along the way that there must be a better design, but I would have argued with
myself that I was close enough to being finished that I might as well continue.

Then I would have tested them and I would have discovered how miserable 56 functions
are to use. I would realize this as I wrote test cases and tried to multiply or add the
matrices using the wrong function. In my less experienced days, I would have persevered
and so have inflicted the bad design on my users. That would not happen these days
because I care about usability. These days, I would have thrown the whole mess away
and started all over again.

All of that would have happened because I had too much invested in a bad design, and the path to the end appeared to be shorter if I just continued what I was doing.

The advantage of designing the code first is that abandoning an approach is so cheap. If you do the design first, you will think to yourself, "There is no way I am going to write fifty-six functions," instead of thinking, "I have so much invested, I may as well continue."

One never becomes so experienced that one can skip the design phase.

17.3.3   Fixing the design

When designs turn bad, go back and imagine yourself as a user of the system. How would you want the function to work? To multiply A and B, I would want to call a function that worked something like this:

```
        whether to transpose A      matrix, regular or sparse
                          \   /
        R = R_multiply(A, tA,  B, tB)
                        /            \
    matrix, regular or sparse            whether to transpose B
```

R_multiply() would return a regular matrix. I would also want another function that returned a sparse matrix but with the same syntax:

```
        whether to transpose A      matrix, regular or sparse
                          \   /
        S = S_multiply(A, tA,  B, tB)
                        /            \
    matrix, regular or sparse            whether to transpose B
```

These two functions would be the only functions SpMat users would need to multiply matrices. As a user, I could type

```
    Result = R_multiply(A,0, B,0)
```

to calculate A*B, or type

```
    Result = R_multiply(A,1, B,0)
```

to calculate A'B, or type

```
    Result = R_multiply(A,0, B,1)
```

to calculate A*B', or type

```
    Result = R_multiply(A,1, B,1)
```

to calculate A'*B'. Regardless of which I typed, A could be a regular or sparse matrix, and so could B. If I wanted a sparse-matrix result, I would type S_multiply() instead of R_multiply().

So I pursued the idea. Could I write the code for the two functions? Notice that
R_multiply() and S_multiply() are not class functions. That is not a problem, but I
admit that it bothered me enough for aesthetic reasons that I thought about recasting
the functions as member functions. I imagined R = A.R_multiply(B, tA, tB) and that
seemed okay, although it bothered me that argument tA was nearer to B than A, so I
imagined R = A.R_multiply(tA, B, tB) instead. That seemed a little better. Then
I realized that A.R_multiply() would work only when A was sparse. When A was a
regular matrix and B was sparse, to multiply A*B I would need another function such
as R = B.R_premultiply(tB, A, tA). And what if A and B were both regular matrices?
I gave up. This solution was just not as elegant as the simple R_multiply() function
that put A and B on equal footing.

I then turned to sketching the code for R = R_multiply(A, tA, B, tB). I wanted to allow
for A and B to be regular or sparse matrices, so I made the arguments transmorphic:

──────────────── Beginnings of a sketch of R_multiply() ────────────

```
real matrix R_multiply(transmorphic A, `boolean´ tA,
                       transmorphic B, `boolean´ tB)
{
        .
        .
        .
}
```

──────────────── Beginnings of a sketch of R_multiply() ────────────

Declaring A and B transmorphic allows them in the door regardless of their types, but
what next? Next I will use eltype() to determine their types so that R_multiply()
can take the appropriate action. You remember eltype(). eltype(M) returns the
element type of M. It returns "real" or "string" or whatever M is. It returns "class"
when M is a class. Mata provides yet another function, classname(M), that returns
the name of the class, such as "SpMat", when eltype(M)=="class". classname(M)
returns "" when eltype(M) is not "class". My sketch of R_multiply() was thus

──────────── Final sketch of R_multiply() using is_sparse() ──────────

```
real matrix R_multiply(transmorphic A, `boolean´ tA,
                       transmorphic B, `boolean´ tB)
{
        A_sparse = is_sparse(A)
        B_sparse = is_sparse(B)

        if (A_sparse) {
                if (B_sparse)  return(R_SxS(A,tA, B,tB))
                else           return(R_SxR(A,tA, B,tB))
        }
        else {
                if (B_Sparse)  return(R_RxS(A,tA, B,tB))
                else           return(R_RxR(A,tA, B,tB))
        }
}
```

──────────── Final sketch of R_multiply() using is_sparse() ──────────

Just as we have done previously, I assumed the existence of functions I would need. R_multiply() is nothing more than a switch that turns the problem over to R_SxS(), R_SxR(), R_RxS(), or R_RxR() as appropriate. In the names, R stands for regular and S stands for sparse. R_RxS() is the subroutine that returns a regular matrix equal to a regular matrix multiplied by a sparse one.

Meanwhile, is_sparse() is a utility routine that I would need to write. I assumed that is_sparse(M) will return true when M is sparse, will return false when M is regular, and will not return at all when M is neither. Here is a sketch of it:

──────────── Sketch of R_multiply() subroutine is_sparse() ──────────

```
`boolean´ is_sparse(transmorphic M)
{
        if (classname(M)=="SpMat")  return(True)
        if (   eltype(M)=="real" )  return(False)
        _error(3250)                // type mismatch error
}
```

──────────── Sketch of R_multiply() subroutine is_sparse() ──────────

We now have sketches of R_multiply() and its utility is_sparse(). And we have a sketch of S_multiply() if we change the names of the R_*x*() subroutines to be S_*x*() and specify that they return a sparse matrix:

──────────── Final sketch of S_multiply() using is_sparse() ────────

```
class SpMat scalar S_multiply(transmorphic A, `boolean´ tA,
                              transmorphic B, `boolean´ tB)
{
        A_sparse = is_sparse(A)
        B_sparse = is_sparse(B)

        if (A_sparse) {
                if (B_sparse) return(S_SxS(A,tA, B,tB))
                else          return(S_SxR(A,tA, B,tB))
        }
        else {
                if (B_Sparse)  return(S_RxS(A,tA, B,tB))
                else           return(S_RxR(A,tA, B,tB))
        }
}
```

──────────── Final sketch of S_multiply() using is_sparse() ────────

And with that, we have eight subroutines left to sketch:

R_SxS()	returns	regular = sparse × sparse
R_SxR()	returns	regular = sparse × regular
R_RxS()	returns	regular = regular × sparse
R_RxR()	returns	regular = regular × regular
S_SxS()	returns	sparse = sparse × sparse
S_SxR()	returns	sparse = sparse × regular
S_RxS()	returns	sparse = regular × sparse
S_RxR()	returns	sparse = regular × regular

17.3.3.1   Sketches of R_*x*() and S_*x*() subroutines

Let's sketch the R_*x*() subroutines first. I have something as simple as this in mind for them:

```
─────────────────────────────── Preliminary sketches of R_*x*() ───────────
real matrix R_OxR(class SpMat scalar A, tA, real matrix B, tB)
{
        return( A.R_SxR(tA,   (tB ? B´ : B)) )
}

real matrix R_RxS(real matrix A, tA, class SpMat scalar B, tB)
{
        return( B.R_RxS( (tA ? A´ : A),   tB) )
}

real matrix R_SxS(class SpMat Scalar A, tA,
                  class SpMat scalar B, tB)
{
        return( A.R_SxS(tA,    B, tB) )
}

real matrix R_RxR(real matrix A, tA, real matrix B, tB)
{
        return( (tA ? A´ : A) * (tB ? B´ : B) )
}
─────────────────────────────── Preliminary sketches of R_*x*() ───────────
```

In the sketch, I assume the existence of class functions that we will write later:

Class function	Multiplies...	By...
A.R_SxR(tA, B)	sparse A or A'	regular B
B.R_RxS(A, tB)	regular A	sparse B or B'
A.R_SxS(tA, B, tB)	sparse A or A'	sparse B or B'

I am imagining that the R_*x*() functions are responsible for transposing regular matrices, if that is called for, and the class functions will be responsible for performing the multiplication and dealing with the possible transposition of the sparse matrices. I do not imagine that the class functions will physically transpose the matrices. I imagine that they will be written to interchange row and column subscripts when necessary, but that is a problem for later.

I am still focusing right now on the sketches of R_*x*() and S_*x*(), and I realized that a simple modification to their code could potentially conserve lots of memory. In the preliminary sketches above, transposition is performed using Mata's in-line transpose operator. For instance, the code for R_SxR() reads

```
real matrix R_SxR(class SpMat scalar A, tA, real matrix B, tB)
{
        return( A.R_SxR(tA, (tB ? B´ : B)) )
}
```

It is the B' in the code that I want to focus on. B' returns the transposition of B. If B is 10,000 × 10,000, then B' is 10,000 × 10,000. That is unfortunate because the new matrix consumes 762 megabytes of memory just as B itself does. Mata provides a function named _transpose() that does not require the extra memory, because coding _transpose(B) replaces the contents of B with B'. _transpose() does not even require that B be symmetric, which is remarkable. Thus, I considered changing the code to use _transpose() even though that would mean using the function twice. When B is changed to contain B's transpose, it will need to be changed back to its original contents before R_SxR() returns.

Thus, I faced the classic conundrum known as the space–time trade-off: consume more space to save time or save the space and consume more time? The only way to decide is to measure the costs and benefits. How much space for how much time? If I had not already known the answer, I would have taken a break from code design and measured the extra time that using _transpose() would cause. We will discuss how to time code in the next chapter. I did know the answer, however. The time cost of _transpose() is small, even when used twice, and I expected the system I was designing would be used by programmers with large matrices. It was an easy decision to make. I changed the sketch of R_SxR() to be

```
real matrix R_SxR(class SpMat scalar A, tA, real matrix B, tB)
{
        real matrix    Result

        if (tB) _transpose(B)
        Result = A.R_SxR(tA, B)
        if (tB) _transpose(B)
        return(Result)
}
```

I changed the sketches of all the R_ *x *() functions:

```
──────────────────────────────── Final sketches of R_*x*() ──────────

real matrix R_SxR(class SpMat scalar A, tA, real matrix B, tB)
{
        real matrix    Result

        if (tB) _transpose(B)
        Result = A.R_SxR(tA, B)
        if (tB) _transpose(B)
        return(Result)
}

real matrix R_RxS(real matrix A, tA, class SpMat scalar B, tB)
{
        real matrix    Result

        if (tA) _transpose(A)
        Result = B.R_RxS(A, tB)
        if (tA) _transpose(A)
        return(Result)
}

real matrix R_SxS(class SpMat Scalar A, tA,
                  class SpMat scalar B, tB)
{
         return( A.R_SxS(tA, B, tB) )
}

real matrix R_RxR(real matrix A, tA, real matrix B, tB)
{
        real matrix    Result

        if (tA) _transpose(A)
        if (tB) _transpose(B)
        Result = A*B
        if (tB) _transpose(B)
        if (tA) _transpose(A)
        return(Result)
}

──────────────────────────────── Final  sketches of R_*x*() ──────────
```

Still left to sketch are the S_*x*() functions. These functions will be nearly identical
to their R_*x*() counterparts. They will call subroutines beginning with S_ instead of
R_, and the functions that they call will return SpMat scalars instead of real matrices.

── Final sketches of S_*x*() ────────

```
class SpMat scalar S_SxR(class SpMat scalar A, tA,
                         real matrix B, tB)
{
        class SpMat scalar  Result

        if (tB) _transpose(B)
        Result = A.S_SxR(tA, B)
        if (tB) _transpose(B)
        return(Result)
}
class SpMat scalar S_RxS(A, tA, class SpMat B, tB)
{
        class SpMat scalar  Result

        if (tA) _transpose(A)
        Result = B.S_RxS(A, tB)
        if (tA) _transpose(A)
        return(Result)
}
class SpMat scalar S_SxS(class SpMat Scalar A, tA,
                         class SpMat scalar B, tB)
{
         return(A.S_SxS(tA,    B, tB))
}
class SpMat scalar S_RxR(A, tA, B, tB)
{
        class SpMat scalar  Result
        if (tA) _transpose(A)
        if (tB) _transpose(B)

        Result.S_RxR(A, B)

        if (tB) _transpose(B)
        if (tA) _transpose(A)

        return(Result)
}
```

── Final sketches of S_*x*() ────────

And with that, all we have left are seven member functions to sketch:

Completed sketch...	calls unsketched class function...
R_multiply(A,tA, B,tB)	
R_SxR(A,tA, B,tB)	A.R_SxR(tA, B)
R_RxS(A,tA, B,tB)	B.R_RxS(A, tB)
R_SxS(A,tA, B,tB)	A.R_SxS(tA, B,tB)
R_RxR(A,tA, B,tB)	
S_multiply(A,tA, B,tB)	
S_SxR(A,tA, B,tB)	A.S_SxR(tA, B)
S_RxS(A,tA, B,tB)	B.S_RxS(A, tB)
S_SxS(A,tA, B,tB)	A.S_SxS(tA, B,tB)
S_RxR(A,tA, B,tB)	Result.S_RxR(A, B)

17.3.3.2   Sketches of class's multiplication functions

When I wrote the linear-regression routines in chapter 11, I used less-accurate but easier-to-program calculation formulas in the first draft. I made the entire system work with the less-accurate functions and later modified the code to be more accurate. That was not pedagogy. I usually focus on building the frame and getting the system working before bothering with better, more complicated algorithms.

I am going to follow the same two-step approach in this project.

The complicating issue in SpMat is efficiently calculating matrix products of sparse matrices. We could start thinking deeply about the issue now, but I suspect that doing so would be counterproductive. First, we have enough design issues to worry about already. Second, second-best routines are easier to program because they are simpler, and we might get lucky and discover that the simple routines work well enough. Third, even if we do not get lucky, we will at least have a working system with certification do-files proving that it works. That means we will be able to easily test the system as we substitute better routines.

We have seven code sketches to write:

Sketch name	For class function...
R=SR	A.R_SxR(tA, B)
R=RS	B.R_RxS(A, tB)
R=SS	A.R_SxS(tA, B,tB)
S=SR	A.S_SxR(tA, B)
S=RS	B.S_RxS(A, tB)
S=SS	A.S_SxS(tA, B,tB)
S=RR	Result.S_RxR(A, B)

I am naming the sketches differently from their class functions because I am not going to bother to sketch the full functions. I am just going to sketch, for instance, R=SS, meaning how to multiply a sparse by a sparse to return a real, and R=SR, meaning how to multiply a sparse by a real, and so on. These routines will be easy to turn into full class functions later.

Not on the list is R=RR, but that is where I am going to start. I want to remind myself, and you, of the general formula for multiplying matrices. It is

$$\mathbf{R} = \mathbf{A} \times \mathbf{B} : \mathbf{R}_{ij} = \sum_{k} \mathbf{A}_{ik}\mathbf{B}_{kj}$$

This formula says that if we were writing a routine to multiply regular matrices and store the result as regular matrix, then the corresponding code to calculate R = A*B would be

```
──────────────────────────────────────────────── Sketch R=RR ──────────
    R = J(rows(A), cols(B), 0)
    for (i=1; i<=rows(A); i++) {
        for (j=1; j<=cols(B); j++) {
            for (k=1; k<=cols(A); k++) {
                R[i,j] = R[i,j] + A[i,k] * B[k,j]
            }
        }
    }
──────────────────────────────────────────────── Sketch R=RR ──────────
```

Modifying sketch R=RR for cases where A is sparse or B is sparse would be easy:

If A is sparse,

Substitute	A.rows()	for	rows(A)
Substitute	A.cols()	for	cols(A)
Substitute	A.get(i,k)	for	A[i,k]

If B is sparse,

Substitute	B.cols()	for	cols(B)
Substitute	B.get(k,j)	for	B[k,j]

With these substitutions, we could write code to produce sketches R=SR, R=RS, and R=SS. I thought about doing that. I thought about modifying R=RR to handle A being sparse. I even wrote out the code and stared at it. I am pretty good with matrices and matrix algebra, and something jumped out to me. I will tell you what it is, but your response might be, "How did he ever notice that?" followed by "And the solution just occurred to him?" When the day comes that you are programming a big system, however, it will concern a subject with which you are familiar, and yes, you will notice things (although perhaps not right ones) and solutions will occur to you (although perhaps not the best ones). We will later discover that I did not notice the right things and neither did the best solutions occur to me, but so it goes. You do your best.

What jumped out to me was that elements of A were accessed multiple times and, in fact, that each row of A would be accessed cols(B)*cols(A) times. I realized that even though I believe A.get(i,j) will return elements reasonably quickly, it will not return them nearly as quickly as Mata would when accessing a regular matrix using its subscript operator, A[i,j]. Performance of the routine would be miserable.

What occurred to me is that the equation for multiplication could be rearranged to access each row once, and then I could get each row once using A.Get(). So I first rewrote the formula $\mathbf{R}_{ij} = \sum \mathbf{A}_{ik}\mathbf{B}_{kj}$ in the "obvious" way:

$$\mathbf{R} = \mathbf{A} \times \mathbf{B} : \mathbf{R}_{i.} = \mathbf{A}_{i.} \times \mathbf{B}$$

where $\mathbf{R}_{i.}$ and $\mathbf{A}_{i.}$ are ith rows of $\mathbf{R}$ and $\mathbf{A}$. Code to implement this solution with sparse matrix A is

────────────────────────── Final sketch R=SR for SpMat::R_SxR() ───────────

```
R = J(A.rows(), cols(B), 0)
 for (i=1; i<=A.rows(); i++) {
        R[i, .] = A.Get(i, .) * B
 }
```

────────────────────────── Final sketch R=SR for SpMat::R_SxR() ───────────

Would this solution run fast enough that I would not have to do further work later? I did not know. All I knew for certain was that the R=SR that I just sketched has to run faster than a direct translation of R=RR modified to work with A being sparse.

I was on a roll. I did not even think about alternative approaches. I solved the problem R=SR by rewriting the formula in row form. I would solve the problem of R=RS by rewriting the formula in column form. In column form, the formula for matrix multiplication is

$$\mathbf{R} = \mathbf{A} \times \mathbf{B} : \mathbf{R}_{\cdot j} = \mathbf{A} \times \mathbf{B}_{\cdot j}$$

where $\mathbf{R}_{\cdot j}$ and $\mathbf{B}_{\cdot j}$ are the jth columns of $\mathbf{R}$ and $\mathbf{B}$. Code to implement this solution with sparse matrix B is

```
——————————————————————— Final sketch R=RS for SpMat::R_RxS() ———————
R = J(A.rows(), cols(B), 0)
for (j=1; j<=B.cols(); j++) {
        R[., j] = A * B.Get(., j)
}

——————————————————————— Final sketch R=RS for SpMat::R_RxS() ———————
```

Finally, if A and B are both sparse, yet another way to write the formula for matrix multiplication is

$$\mathbf{R} = \mathbf{A} \times \mathbf{B} : \mathbf{R}_{ij} = \mathbf{A}_{i \cdot} \times \mathbf{B}_{\cdot j}$$

where $\mathbf{A}_{i \cdot}$ is the ith row of $\mathbf{A}$, and $\mathbf{B}_{\cdot j}$ is the jth column of $\mathbf{B}$. The corresponding code is

```
——————————————————————— Final sketch R=SS for SpMat::R_SxS() ———————
R = J(A.rows(), cols(B), 0)
for (i=1; i<=A.rows(); i++) {
        A_i = A.Get(i, .)
        for (j=1; j<=B.cols(); j++) {
                R[i, j] = A_i * B.Get(., j)
        }
}

——————————————————————— Final sketch R=SS for SpMat::R_SxS() ———————
```

In this code, sadly, each column of B will be accessed A.rows() times, so I was fearful that the performance of R=SS will be worse than that of R=SR and R=RS.

The sketches for R=SR, R=RS, and R=SS can easily be converted into sketches for S=SR, S=RS, and S=SS. Each is a line-by-line translation of the corresponding R=... routine. For instance, to produce the sketch of S=SR, I started with R=SR,

```
R = J(A.rows(), cols(B), 0)
for (i=1; i<=A.rows(); i++) {
        R[i, .] = A.Get(i, .) * B
}
```

and I translated it to produce

```
S.setup(A.rows(), cols(B))
for (i=1; i<=A.rows(); i++) {
        S.Put(i,. , A.Get(i, .) * B)
}
```

The first line creates matrix R (S) to be `A.rows()` × `cols(B)` containing 0s. Built-in function `J()` does that in R=SR. Function `S.setup()` does the same in S=SR.

The second line is the same in both R=SR and S=SR.

The third line puts `A.Get(i,.)*B` in the ith row of R (S). Here are the two lines listed one after the other:

```
R[i, .] =   A.Get(i, .) * B
S.Put(i,. , A.Get(i, .) * B)
```

I translated the other S=... routines the same way. Here are all but one of them:

──────────────────── Final sketch S=SR for SpMat::S_SxR() ────────

```
S.setup(A.rows(), cols(B))
for (i=1; i<=A.rows(); i++) {
        S.Put(i,. , A.Get(i, .) * B)
}
```

──────────────────── Final sketch S=SR for SpMat::S_SxR() ────────

──────────────────── Final sketch S=RS for SpMat::S_RxS() ────────

```
S.setup(rows(A), B.cols)
for (j=1; i<=B.cols(); j++) {
        S.put(.,j , A * B.Get(., j))
}
```

──────────────────── Final sketch S=RS for SpMat::S_RxS() ────────

──────────────────── Final sketch S=SS for SpMat::S_SxS() ────────

```
S.setup(A.rows(), B.cols())
for (i=1; i<=A.rows(); i++) {
        A_i = A.Get(i, .)
        for (j=1; j<=B.cols(); j++) {
                R.put(i,j, A_i * B.Get(., j))
        }
}
```

──────────────────── Final sketch S=SS for SpMat::S_SxS() ────────

We set out to write sketches for seven functions and have written six:

To be written	Sketch
A.R_SxR(tA, B)	R=SR (done)
A.S_SxR(tA, B)	S=SR (done)
B.R_RxS(A, tB)	R=RS (done)
B.R_RxS(A, tB)	S=RS (done)
A.R_SxS(tA, B, tB)	R=SS (done)
A.S_SxS(tA, B, tB)	S=SS (done)
Result.S_RxR(A, tA, B, tB)	S=RR

The sketch for S=RR could be as simple as two lines of code, namely,

```
Result.setup(rows(A), cols(B))
Result.Put(., ., A*B)
```

Because the caller is requesting the result be stored as sparse, it would perhaps be better to calculate the result row by row and so reduce the amount of memory used:

```
——————————————————————— Final sketch S=RR for SpMat::S_RxR() ———————
Result.setup(rows(A), cols(B))
for (i=1; i<=rows(A); i++) {
        Result.Put(i, .) = A[i,.] * B
}
——————————————————————— Final sketch S=RR for SpMat::S_RxR() ———————
```

None of the above sketches handle arguments tA and tB. Those arguments, I remind you, are instructions to treat A and B as if they were transposed. That is a detail we can add when we write the final code. If a sketch requires A.Get(i, .), for instance, then when tA is true, what is needed is A.Get(., i)'.

17.3.4 Design summary

We have completed the design.

"That is a design?" I can almost hear you thinking. I admit that there are still lots of problems left unsolved, but I believe that solving them will be no more difficult than the usual problems we encounter when we write code. The purpose of a design is to provide the solutions to the big problems and to convince ourselves that we could solve whatever problems remain. There are many issues that we have ignored in the design. Handling tA and tB in our code sketches are two issues that we happen to know about, but we will discover others. We hope those issues do not invalidate our design, but I cannot guarantee that.

When you do your own design for a project, you will write notes similar to those I have written above. Your own design notes will not include the lengthy explanations that I included. I included them to explain my thinking to you. Real design notes are notes to yourself and include only the explanations you need so that your future self can understand them.

Your notes will share the same stream-of-consciousness flavor of the notes I have shown you. You have seen me sketch a function and resketch it after further consideration. You will have such repeated sketches in your own notes. I do not recommend amending or editing them. I do not edit mine because I sometimes need to fall back to a previous thought or idea.

Anyway, you just decide at some point that you have done enough design and that it is time to start writing code. But before you do that, you should make a Design Summary from your notes. My Design Summaries are often just a list of functions to be implemented, but that can vary. A list of functions will be sufficient in this case. Table 17.1 below is my Design Summary just as I wrote it for myself. The summary table has three columns: Function name, Exposure, and Returns.

The Exposure column is filled in with the terms public, private, or DND (do not document). DND functions are functions that I wish could be private but cannot be. I would have preferred that the `A.R_*x*()` and `A.S_*x*()` functions be private, but they cannot be because they are called by the subroutines of the outside-the-class functions `R_multiply()` and `S_multiply()`.

The Returns column is what the function returns. Regular and Sparse mean regular and sparse matrix, meaning real matrix and class `SpMat` scalar.

The Design Summary is short, but you know how much work went into it, and even more work went into it than I told you about. I renamed the functions before you ever saw them. The names in the table are the names we have been using throughout this chapter, but they are not the names that were in my original notes to myself. I seldom get function names right at the outset, and getting them right is important. Function names should describe what the function does.

Table 17.1. Design Summary

Function name	Exposure	Returns
A.setup(rows, cols)	public	void
A.rows()	public	real scalar
A.cols()	public	real scalar
A.put(i, j, v)	public	void
A.replace(i, j, v)	public	void
A.get(i, j)	public	real scalar
A.Put(i, j, |i1, j1)	public	void
A.Replace(i, j, |i1, j1)	public	void
A.Get(i, j, |i1, j1)	public	real matrix
R_multiply(A, tA, B, tB)	public	Regular
S_multiply(A, tA, B, tB)	public	Sparse
R_SxS(A, tA, B, tB)	DND	Regular
R_SxR(A, tA, B, tB)	DND	Regular
R_RxS(A, tA, B, tB)	DND	Regular
R_RxR(A, tA, B, tB)	DND	Regular
S_SxS(A, tA, B, tB)	DND	Sparse
S_SxR(A, tA, B, tB)	DND	Sparse
S_RxS(A, tA, B, tB)	DND	Sparse
S_SxR(A, tA, B, tB)	DND	Sparse
A.R_SxS(tA, B, tB)	DND	Regular
A.S_SxS(tA, B, tB)	DND	Sparse
A.R_SxR(tA, B, tB)	DND	Regular
A.S_SxS(tA, B, tB)	DND	Sparse
A.R_RxS(tA, B, tB)	DND	Regular
A.S_RxS(tA, B, tB)	DND	Sparse
R.S_SxS(A, tA, B, tB)	DND	Sparse

After you have put the functions in the summary table in logical order, it is an excellent time to rename them. I recommended earlier that you not edit your notes, but do edit them when you change function names. I do. In my original notes, the functions that you know as A.R_RxS() and A.R_SxR() started life as A.S_premultiply_R() and A.R_postmultiply_S(). After constructing the table, I changed them to A.R__R_x_this() and A.R__this_x_R(), but I did not much like those names either. I finally stumbled on the names that you know, A.R_RxS() and A.R_SxR().

I also did more design work than I told you about. I designed functions for matrix addition, too. You will be able to understand the design just from the function names in table 17.2:

Table 17.2. Design Summary, continued

Function name	Exposure	Returns
R_add(A, tA, B, tB)	public	Regular
S_add(A, tA, B, tB)	public	Sparse
R_SpS(A, tA, B, tB)	DND	Regular
R_SpR(A, tA, B, tB)	DND	Regular
R_RpS(A, tA, B, tB)	DND	Regular
S_SpS(A, tA, B, tB)	DND	Sparse
S_SpR(A, tA, B, tB)	DND	Sparse
S_RpS(A, tA, B, tB)	DND	Sparse
A.R_SpS(tA, B, tB)	DND	Regular
A.S_SpS(tA, B, tB)	DND	Sparse
A.R_SpR(tA, B)	DND	Regular
A.S_SpR(tA, B)	DND	Sparse
R.S_RpR(A, B)	DND	void

17.3.5   Design shortcomings

You now have two documents: Design Notes and Design Summary. The third document you should prepare is Design Shortcomings. This is your last chance to stop and reconsider before writing code. The purpose of Design Shortcomings is to force you to think about said shortcomings and whether you really are ready to move to writing code.

Here is my original Design Shortcomings document:

——————————————————————— Design Shortcomings document ————

Right now, the system handles sparse matrices. For performance reasons, I suspect that we should include special cases such as symmetric-and-sparse, diagonal, and banded. If necessary, they can be added later. Right now, I will make sure that all sparse-matrix access is handled by just a few functions. Thus, if I need to add other storage formats, I can modify them to make things work, although I suspect I will want to write custom multipliers and adders to achieve better performance.

I have not thought deeply about matrix multiplication. What I have sketched will work but may have performance issues.

I have not thought deeply about matrix addition. What I have sketched will work but may have performance issues.

What about matrix subtraction?
What about matrix transposition?
What about matrix inversion and solvers?

Subtraction: Will work like addition but will need two extra functions because subtraction is not commutative. The extra functions will be `A.R_RmS()` and `A.S_RmS()`.

Transposition: We need `S.transpose()` to return a transposed matrix. It should be easy to write.

Inversion: There is a partitioned inverse formula. I suspect we can use it. If the matrix is really big, the code may need to recurse on itself. Adding inversion is, I suspect, a moderately difficult problem.

——————————————————————— Design Shortcomings document ————

Nothing in my shortcomings document makes me think we should not proceed.

17.4   Code

The code starts the way all `.mata` files start, with the opening lines

——————————————————— extract from spmat1.mata ————
```
*! version 1.0.0
version 15

set matastrict on
```
——————————————————— extract from spmat1.mata ————

Next in the file, I define macros, the derived types, that I will use

```
────────────────────────────────── extract from spmat1.mata ──────────
local Sp             SpMat

local RS             real scalar
local RR             real rowvector
local RC             real colvector
local RM             real matrix
local Sparse         class `Sp' scalar
local Regular        real matrix
local AA             class AssociativeArray scalar

local boolean        real scalar
      local True   1
      local False  0
──────────────────────────────────── extract from spmat1.mata ──────────
```

'Sp' stands for SpMat, which is the development name I will use for the class. We discussed naming issues in sections 14.4.7 and 12.5. If I decide to continue this work to produce an official StataCorp class for release to users, I will change 'Sp' to stand for SparseMatrix. At StataCorp, we have decided that official StataCorp-provided class names will be fully spelled out with words, each starting with a capital letter. I recommend that you give your classes abbreviated names, such as SpMat—although because I just took that name, you should find another one.

'RS', ..., 'RM' mean the usual thing.

I created 'Sparse' and 'Regular' to designate sparse and regular matrices. The terms appear throughout my design document, and I decided to preserve that in the code.

'AA' stands for class AssociativeArray scalar, the StataCorp-provided class described in chapter 16. StataCorp may use long names, but I do not like typing them.

I created type 'boolean' to contain 'True' and 'False'. I often call this type 'booleanS'—the S on the end to emphasize that it is a scalar. I may regret not having called the type 'booleanS' in this project if I later need a boolean vector, but I do prefer the shorter name.

The next line enters Mata:

```
────────────────────────────────── extract from spmat1.mata ──────────
mata:
──────────────────────────────────── extract from spmat1.mata ──────────
```

I will not exit Mata until the last line of the file, some hundreds of lines away.

Inside the `mata:...end` block, I first define the class:

```
                                              extract from  spmat1.mata
class `Sp´
{
    public:
        void                setup()

        `RS´                rows(), cols()

        `RS´                get()
        void                put(), replace()

        `RM´                Get()
        void                Put(), Replace()

        //      returns = A*B          R = Regular matrix
        //             \ | |           S = Sparse  matrix
        `Regular´       R_SxR()
        `Sparse´        S_SxR()
        `Regular´       R_RxS()
        `Sparse´        S_RxS()
        `Regular´       R_SxS()
        `Sparse´        S_SxS()
        void            S_RxR()

        //      returns = A+B          R = Regular matrix
        //             \ | |           S = Sparse  matrix
        `Regular´       R_SpS()
        `Sparse´        S_SpS()
        `Regular´       R_SpR()
        `Sparse´        S_SpR()
        void            S_RpR()

    private:
        `RS´                rowst(), colst()

        `RM´                Get2(),      Get4()
         void               Put3(),      Put5()
         void               Replace3(), Replace5()

        void                set_row(),   set_col()

        `AA´                AA
        `RS´                rows, cols
}
                                              extract from  spmat1.mata
```

I want to underscore how neatly formatted and easy to read the class definition is. Notice that I use comments to explain my function naming convention.

Those of you from a scientific research background may be used to less neatly formatted files. The contents of do-files that record scientific research are, in fact, usually a mess. Nonetheless, the files serve their purpose. They can be rerun to re-create scientific results, and that is what is important. That the files are messy merely means that you or anyone else needing to understand the code will have to work harder. The intent of the research is usually laid out in professional papers, so even with a messy file, it is possible to verify that the code matches its published intent.

There will be no professional paper backing up this code. If this code is successful, meaning distributed and used by others, it is a near certainty that the code will need to be modified someday. It will save everyone time if the code is neatly formatted and cleanly arranged. Neatly formatting your code will save even your time in the future.

I next defined the opening functions described in the design:

```
———————————————————————————————— extract from  spmat1.mata ————
void `Sp´::setup(`RS´ rows, `RS´ cols)
{
        AA.reinit("real", 2)
        AA.notfound(0)

        this.rows = rows
        this.cols = cols
}

`RS´ `Sp´::rows()                       return(rows)
`RS´ `Sp´::cols()                       return(cols)

`RS´ `Sp´::get(`RS´ i, `RS´ j)     return( AA.get((i,j)) )

void `Sp´::put(`RS´ i, `RS´ j, `RS´ v)
{
        if (v) AA.put((i,j), v)
}

void `Sp´::replace(`RS´ i, `RS´ j, `RS´ v)
{
        if (v)     AA.put( i, j, v)
        else    AA.remove((i, j))
}
—————————————————————————————— extract from  spmat1.mata ————
```

The above functions are exactly as I outlined them in the design document.

I next wrote Get(), one of the overloaded, capital-letter functions with so many features. "Overloaded" in this case simply means that the function does different things depending on the number of arguments specified. Just because the function is overloaded does not mean the code has to be. The purpose of Get() is simply to route control to the appropriate subroutine.

```
———————————————————————————————— extract from  spmat1.mata ————
`RM´ `Sp´::Get(one, two, |three, four)
{
        if (args()==2) return(Get2(one, two))
        if (args()==4) return(Get4(one, two, three, four))
        else           _error(3001)   // incorrect # of args
}
—————————————————————————————— extract from  spmat1.mata ————
```

If `Get()` is called with two arguments, control is transferred to `Get2()`. If there are four arguments, control is transferred to `Get4()`.

Notice that I named the arguments **one**, **two**, **three**, and **four**. At this point, I am not thinking substantively about what `Get()` does; I am thinking merely that it has a two-argument form and a four-argument form, and I am dealing with that. I let all four arguments default to transmorphic. The required types will be declared and checked later, by `Get2()` or `Get4()`.

Functions `Get2()` and `Get4()` appear later in the file. Before defining them, I went ahead and defined the other two overloaded, capital-letter functions:

────────────────────────────────────── extract from `spmat1.mata` ─────────
```
void `Sp´::Put(one, two, three, | four, five)
{
        if       (args()==3)   Put3(one, two, three)
        else if  (args()==5)   Put5(one, two, three, four, five)
        else     _error(3001)  // incorrect # of args
}

void `Sp´::Replace(one, two, three, | four, five)
{
        if       (args()==3) Replace3(one, two, three)
        else if  (args()==5) Replace5(one, two, three, four, five)
        else     _error(3001)  // incorrect # of args
}
```
────────────────────────────────────── extract from `spmat1.mata` ─────────

I then wrote `Get2()` and `Get4()`, which means I turned to thinking substantively. `Get2()` handles the syntaxes

scalar = S.Get(i, j)	same as S.get()
rowvector = S.Get(i, .)	get row
colvector = S.Get(., j)	get column
matrix = S.Get(., .)	get matrix

`Get4()` handles the syntax

submat = S.Get(i1,j1, i2,j2)	submat = S[|i1,j1 i2,j2|]

In the following code, first notice that the arguments are now declared with explicit types. Next notice that `Get2()` calls `get()` or `Get4()` with the appropriate arguments.

──────────────────────────────────── extract from spmat1.mata ────────

```
`RM´ `Sp´::Get2(`RS´ i, `RS´ j)
{
        if (i==. && j==.) return(Get4(1,1, rows, cols))
        if (        j==.) return(Get4(i,1,    i, cols))
        if (i==.)         return(Get4(1,j, rows,    j))
        return(get(i,j))
}

`RM´ `Sp´::Get4(`RS´ r0, `RS´ c0, `RS´ r1, `RS´ c1)
{
        `RS´      r, c   // all r*, c* vars in this metric
        `RS´      i, j   // i, j              in R[]  metric
        `RM´      R

        if (r0<1     | r0>r1)   _error(3300)   // arg out of range
        if (c0<1     | c0>c1)   _error(3300)
        if (r1>rows | c1>cols) _error(3300)

        R = J(r1-r0+1, c1-c0+1, 0)

        i = 0
        for (r=r0; r<=r1; r++) {
                ++i
                j = 0
                for (c=c0; c<=c1; c++) R[i, ++j] = get(r, c)
        }

        return(R)
}
```

──────────────────────────────────── extract from spmat1.mata ────────

Put3(), Put5(), Replace3(), and Replace5() are written in the same style as Get2()
and Get4().

```
──────────────────────────────────────── extract from spmat1.mata ────────
void `Sp´::Put3(`RS´ i, `RS´ j, `RM´ R)
{
        if       (i==. && j==.) Put5(1,1, rows,cols, R)
        else if (        j==.) Put5(i,1,    i,cols, R)
        else if (i==.)         Put5(1,j, rows,    j, R)
        else                   put(i, j, R)
}

void `Sp´::Put5(`RS´ r0, `RS´ c0, `RS´ r1, `RS´ c1, `RM´ R)
{
        `RS´    i, j
        `RS´    r, c

        i = 0
        for (r=r0; r<=r1; r++) {
                ++i
                j = 0
                for (c=c0; c<=c1; c++) put(r, c, R[i,++j])
        }
}

void `Sp´::Replace3(`RS´ i, `RS´ j, `RM´ R)
{
        if (i==. && j==.) Replace5(1,1, rows,cols, R)
        else if (j==.)    Replace5(i,1,    i,cols, R)
        else if (i==.)    Replace5(1,j, rows,    j, R)
        else              replace(i, j, R)
}

void `Sp´::Replace5(`RS´ r0, `RS´ c0, `RS´ r1, `RS´ c1, `RM´ R)
{
        `RS´    i, j
        `RS´    r, c

        i = 0
        for (r=r0; r<=r1; r++) {
                ++i
                j = 0
                for (c=c0; c<=c1; c++) replace(r, c, R[i,++j])
        }
}
──────────────────────────────────────── extract from spmat1.mata ────────
```

I have written functions in the order specified in the Design Summary so far. The
summary says that outside-the-class functions R_multiply() and S_multiply() are to
be written next, but I decided I would write the class multiplication subroutines that
those two functions will call.

I started with R_SxR():

```
——————————————————————————————— extract from spmat1.mata ——————————
`Regular´ `Sp´::R_SxR(`boolean´ tA, `Regular´ B)
{
        `Regular´  R
        `RR´       A_i
        `RS´       rA, cA, rB, cB, i

        // R = A*B where A=this (sparse)

        rA = (tA ? cols : rows)
        cA = (tA ? rows : cols)
        rB = (::rows(B))
        cB = (::cols(B))

        if (cA != rB) _error(3200)     // conformability error

        R   = J(rA, cB, 0)
        A_i = J( 1, cA, 0)
        for (i=1; i<=rA; i++) {
                set_row(A_i, i, tA)
                R[i, .] = A_i * B
        }
        return(R)
}
—————————————————————————————— extract from spmat1.mata ——————————
```

The above code corresponds to the sketch

```
————————————————————————————————————————————— R=SR ——————————
R = J(A.rows(), cols(B), 0)
for (i=1; i<=A.rows(); i++) {
        R[i,.] = A.Get(i, .) * B
}
————————————————————————————————————————————— R=SR ——————————
```

The code appears to differ from the sketch more than it really does. It looks so different because the final code handles details such as tA that the sketch did not. It also looks different because of the syntax of class programs. I wrote the sketch as if I were writing an outside-the-class routine. Thus, the sketch refers to A.rows() and A.cols(), which become simply rows and cols in the code. The code could have referred to this.rows and this.cols had I wanted to emphasize that they are class member variables. Meanwhile, the sketch uses cols(B), and that became ::cols(B) in the code because the class itself defines a cols() function.

I also made a substantive change. Compare the loop in the final code with the loop in the sketch.

Final code:

```
for (i=1; i<=rA; i++) {
        set_row(A_i, i, tA)
        R[i, .] = A_i * B
}
```

Sketch (translated to class style):

```
for (i=1; i<=rows(); i++) {
        R[i,.] = Get(i, .) * B
}
```

In the code, I wrote a private class function named **set_row()** to perform the actions of **Get()** that I used in the sketch. I did this in part to handle the transposition issue. The syntax of **set_row()** is

```
(void) set_row(A_i, i, tA)
```

The function returns in **A_i** the **i**th row of the matrix if **tA** is 0, or it returns the **i**th column of the matrix, transposed. Function **set_row()** appears later in **spmat1.mata**, but let me show it to you now:

```
──────────────────────────────── extract from spmat1.mata ─────────
void `Sp'::set_row(`RR' row_i, `boolean' transposed, `RS' i)
{
        `RS'  j, cA

        cA = (::cols(row_i))

        if (transposed) {
                for (j=1; j<=cA; j++) row_i[j] = get(j, i)
        }
        else {
                for (j=1; j<=cA; j++) row_i[j] = get(i, j)
        }
}
──────────────────────────────── extract from spmat1.mata ─────────
```

And with that, I have nothing more to say about the code except to recommend that you go through it. You can see the full listing of the **SpMat** version-1 code by typing in Stata

```
. view ~/matabook/spmat1.mata
```

If you have not yet downloaded the files associated with this book, see section 1.4.

17.5   Certification script

Of course I wrote a certification script. I always write a certification script. And this certification script, just like all certification scripts, begins with the lines

```
─────────────────────────────────────── test_spmat1.do ───────────
// version intentionally omitted
clear all
run spmat1.mata

    .
    .
─────────────────────────────────────── test_spmat1.do ───────────
```

If you have downloaded the files associated with this book as described in section 1.4, you can see the final version of the certification script by typing in Stata

```
. view ~/matabook/test_spmat1.do
```

Read my comments here first, however.

I wrote the code and test script simultaneously, in fact. I printed the Design Notes and Design Summary and put them next to me. Meanwhile, I opened three windows on my computer's screen. I edited **spmat1.mata**, the source code, in one. In another, I edited **test_spmat1.do**, the test script. In the third, I opened Stata. Thus, I could write code, add to the test script, and run the test script just by tabbing through windows.

The original **test_spmat1.do** file contained the line **do spmat1.mata** rather than **run spmat1.mata** so that I could see the output from compiling the code. If there was an error—and there were lots of them—I could tab over to the **spmat1.mata** file, fix it, tab back to Stata, and run **test_spmat1.do** again. I do not move from writing one routine to the next until the compile-time bugs are removed.

I test routines as I go when that is possible. The tests are not thorough, but I test enough to verify that the routine works in at least one case. Once I finish writing the code, I add more tests to the end of the script. If you look carefully at **test_spmat1.do**—but wait to do that—you will see that I tested every combination of sparse and regular matrices.

I also recommend testing code on nonsquare matrices whenever that is possible. When writing code, it is easy to code **rows** when you mean **cols** and vice versa. If you test with square matrices, you will never uncover those mistakes.

Now look at the certification script **test_spmat1.do**.

The **spmat1.mata** code passes certification, but I did not add **spmat1.mata** to the **lmatabook.mlib** library because we are not yet done with development. We know that the current code has second-rate multiplication routines. The routines produce correct results—the certification script establishes that—but we still need to measure how quickly the routines run. If performance is not adequate, we will need to develop better routines. That is the topic of the next chapter.

18 Programming example: Sparse matrices, continued

18.1 Introduction

We are partway through the development process. We set about producing a class to provide efficient handling of sparse matrices in the previous chapter. We designed the system and produced three documents: Design Notes, Design Summary, and Design Shortcomings. Our design was incomplete—all designs are—but we judged it to be complete enough that we could proceed to write code. We have written code and a certification do-file. We now have a known-to-be-working system.

Now we need to evaluate it. We implemented second-rate algorithms for matrix multiplication, and those algorithms may not run quickly enough. This is much like the situation we found ourselves in back in chapter 11 when we developed the linear-regression system. Back then, we used second-rate numerical algorithms and needed to evaluate the system on the basis of accuracy. Even if we had started with first-rate algorithms, we would have needed to evaluate the system just to prove that we had been successful.

We evaluated the previous system on the basis of accuracy. We will evaluate this system on the basis of run times. We will do timings to evaluate the performance of SpMat. Here is what we will discover:

Time to multiply 1000×1000 banded matrices				
Timing	Method		Types of matrices	Execution time
T1	Mata	A*B	R=RR	0.0081 seconds
T2	SpMat	R_multiply()	R=RR	0.0081 seconds
T3	SpMat	R_multiply()	R=SR	2.873 seconds
T4	SpMat	R_multiply()	R=RS	3.339 seconds
T5	SpMat	R_multiply()	R=SS	1,266.68 seconds

R means regularly stored matrix (real matrix).
S means sparsely stored matrix (class SpMat scalar).

Timing T1 reveals that Mata itself multiplies regular 1000×1000 matrices in 0.0081 seconds. I made T1 for use as a benchmark.

Timing T2 reveals that the R_multiply() function we wrote multiplies regular matrices just as quickly as Mata does. I was pleased but not surprised. R_multiply() uses Mata to multiply regular matrices.

Timings T3 and T4 show that R_multiply() takes longer to multiply sparse and regular matrices, but that was to be expected. R_multiply() needs 2.873 seconds or 3.339 seconds depending on whether we multiply sparse times regular or regular times sparse. The timings are disappointing but performance is perhaps tolerable.

Meanwhile, timing T5 reveals that R_multiply() needed 1,266.68 seconds (21 minutes!) to multiply sparse with sparse matrices. Twenty-one minutes is unacceptable.

We obviously need to develop a faster algorithm for R=SS, and perhaps faster algorithms for R=SR and R=RS, too.

Before we do that, I will show you how I obtained these timings. I also will show you how to make detailed timings, meaning timings of parts of the code, where the parts can be loops or even individual lines of code. You perform detailed timings in hopes of finding performance bottlenecks that you can fix. Bottlenecks come in all sizes, but they seldom exceed 50%. It might be reasonable for us to have such hopes for R=SR and R=RS, but even a 50% improvement would not be reasonable to save R=SS.

And finally, we will develop new algorithms to make the system run faster.

We begin by making the overall timings previewed in the above table.

18.2   Making overall timings

To perform overall timings, you set Stata's **rmsg** on and run examples. You set **rmsg** on by typing **set rmsg on** in Stata, not Mata.

```
. set rmsg on
r; t=0.00 14:07:18

. _
```

With **rmsg** on, Stata reports the time it takes to run the commands you type, such as

```
. quietly regress y x1 x2 x3 x4
r; t=0.06 14:07:18

. _
```

The **t=0.06** in the return message means that **regress** ran in 0.06 seconds. The timing is measured from the instant I pressed *Enter* until the command completed. I typed **quietly** in front of **regress** not just to save paper but also so that the timing would not be contaminated by the time it takes to display the output on the screen. The **r;** line also reports the current time of day. I ran the timing at 14:07 (2:07 p.m.). I will delete the time of day from subsequent output.

Setting **rmsg** on causes Stata but not Mata to show timings. Even so, we can obtain timings for Mata code by running it from the Stata prompt using Stata's **mata:** prefix command. We could time multiplication of regular matrices by typing

```
. mata: R = A*B
r; t=0.01
```

R = A*B runs in 0.01 seconds. When I ran this example, matrices **A** and **B** were each 1000×1000 banded matrices with 1s along the diagonal and 0.5s on either side of it. I defined **A** and **B** and then ran the command I wanted to time. I wrote **makemat()** to create the matrices:

```
. mata: A = makemat(1000)
r; t=0.00

. mata: B = makemat(1000)
r; t=0.00

. mata: R = A*B
r; t=0.02
```

I will show you the `makemat()` code in a moment.

Notice that `mata: R = A*B` took 0.02 seconds instead of 0.01 this time. Variation in run times is common. Timings vary randomly from one run to the next for lots of reasons. If the computer is being shared with others, that can cause timings to vary, although my computer is private. My computer is connected to the Internet, however, and that causes timings to vary as messages fly by and demand a little of the computer's time. Computers, yours and mine, also run background tasks just to maintain the environment. Those tasks continually awake, perform their duties, and fall back to sleep.

This makes short timings suspect. When a function runs too quickly, you can obtain more-accurate results by running the function repeatedly and calculating the average time. For instance, I could run `R = A*B` 100 times by typing

```
. mata: for (i=1; i<=100; i++) R = A*B
r; t=0.81
```

The more-accurate measurement is $0.81/100 = 0.0081$ seconds for multiplying `A` and `B`. The measurement is a slight overestimate because it includes the time to run the `for` loop, but the loop runs so quickly that we can ignore it. I separately timed the `for` loop and discovered that it runs in roughly 0.000007 seconds on my computer.

We will use `rmsg` with looping to perform five timings:

Timing	Case	Mata code
T1	Mata R=RR	R = A*B
T2	SpMat R=RR	R = R_multiply(A,0, B,0)
T3	SpMat R=SR	R = R_multiply(SA,0, B,0)
T4	SpMat R=RS	R = R_multiply(A,0, SB,0)
T5	SpMat R=SS	R = R_multiply(SA,0, SB,0)

A and B will be 1000×1000 regular matrices.
SA and SB will be 1000×1000 sparse matrices.

The matrices will all be the same size and contain the same values. Each will be 1000×1000 and have 1s along the diagonal and 0.5s on either side of it. A and B will be stored as regular matrices, and SA and SB will be stored as SpMats. I made a do-file to construct the matrices:

```
──────────────────────────────────── timingsetup.do ─────────

// version number omitted because of laziness.
// It will not matter because I will be throwing
// timingsetup.do away soon.
clear all
run spmat1.mata

mata:
real matrix makemat(real scalar N)
{
        real matrix    R
        real scalar    i

        R = J(N, N, 0)
        R[1,1] = 1
        for (i=2; i<=N; i++) {
                R[i,i] = 1
                R[i-1,i] = R[i,i-1] = 0.5
        }
        return(R)
}
A = makemat(1000)
B = makemat(1000)

SA = SpMat()
SA.setup(1000, 1000)
SA.Put(., ., A)

SB = SpMat()
SA.setup(1000, 1000)
SA.Put(., ., B)
──────────────────────────────────── timingsetup.do ─────────
```

18.2.1 Timing T1, Mata R=RR

I produced timing T1 by typing do timingsetup followed by

```
. mata: for (i=1; i<=100; i++)  R = A*B
r; t=0.81
```

Thus, T1 = 0.81/100 = 0.0081 seconds.

18.2.2 Timing T2, SpMat R=RR

I produced timing T2 by typing

```
. mata: for (i=1; i<=100; i++)  R = R_multiply(A, 0, B, 0)
r; t=0.81
```

Thus, T2 = 0.81/100 = 0.0081 seconds.

I expected that T1 would equal T2. To remind you, `R_multiply()` calls `R_RxR()`, and `R_RxR()` is

```
———————————————————————————— extract from spmat1.mata ————————
`Regular´ R_RxR(`Regular´ A, `boolean´ tA,
                `Regular´ B, `boolean´ tB)
{
        `Regular´  R

        if (tA) _transpose(A)
        if (tB) _transpose(B)
        R = A*B
        if (tB) _transpose(B)
        if (tA) _transpose(A)

        return(R)
}
——————————————————————————————— extract from spmat1.mata ————————
```

18.2.3   Timing T3, SpMat R=SR

I produced timing T3 by typing

```
. mata: for (i=1; i<=100; i++) R = R_multiply(SA, 0, B, 0)
—Break—
r(1); t=59.22
```

Well, that did not work, although I waited almost a minute before pressing *Break*. Then I tried again but performed the command only 10 times:

```
. mata: for (i=1; i<=10; i++) R = R_multiply(SA,0,  B,0)
r; t=28.73
```

Thus, T3 = 28.73/10 = 2.873 seconds.

It is disappointing that `R=SR` runs $2.873/0.0081 = 355$ times slower than `R=RR`. We will later do detailed timings to find out why.

18.2.4   Timing T4, SpMat R=RS

I expected timing T4 to equal timing T3. T3 timed a sparse times a regular, whereas T4 reverses the order. Given the experience with timing T3, I ran timing T4 only 10 times:

```
. mata: for (i=1; i<10; i++) R = R_multiply( A,0,  SB,0)
r; t=33.39
```

Thus, T4 = 33.39/100 = 3.339 seconds.

The 3.339 seconds surprised me because T3 ran in just 2.873 seconds. I was so surprised by the 16% difference that I ran T3 and T4 a few more times. Results were unchanged, so that is one more thing we will need to look into later.

18.2.5 Timing T5, SpMat R=SS

Timing T5 is the time to multiply two sparse matrices. I tried running it 10 times but got tired of waiting. Then I ran it just once:

```
. mata: R = R_multiply(SA,0, SB,0)
r; t=1266.68 15:10:52
```

The 1,266.68 seconds is over 21 minutes!

I felt obligated to wait all 21 minutes so that I could show you the exact result, but I went back and made the exact timing later. When I ran T5 for the first time, I pressed *Break* after waiting about a minute, and then I ran timings on smaller matrices and used those results to predict the time for multiplying 1000×1000 matrices. My prediction was roughly 20 minutes.

18.2.6 Call a function once before timing

When you run your own timings, you should usually execute the function once, ignore the timing, and then make the timing. You should do that because Mata may need to load the function or its subroutines from Mata libraries the first time you run it.

In the timings I made, the code being timed was in **spmat1.mata**, and that file was loaded when I typed **do timingsetup**. Thus, only the odd Mata library function might have needed to be found and loaded, and I ignored the issue.

18.2.7   Summary

You have already seen the table, but to remind you, here it is again:

Time to multiply 1000×1000 banded matrices				
Timing	Method		Types of matrices	Execution time
T1	Mata	A*B	R=RR	0.0081 seconds
T2	SpMat	R_multiply()	R=RR	0.0081 seconds
T3	SpMat	R_multiply()	R=SR	2.873 seconds
T4	SpMat	R_multiply()	R=RS	3.339 seconds
T5	SpMat	R_multiply()	R=SS	1,266.68 seconds

R means regularly stored matrix (`real matrix`).
S means sparsely stored matrix (`class SpMat scalar`).

As I said, the performance of R=SS is obviously unacceptable, the performance of R=SR and R=RS is disappointing, and the difference in run times between R=SR and R=RS is unexpected.

Compared with the run time for R=SS, the other timings pale in importance. If the 21 minutes R=SS needed cannot be reduced, it will call into question the entire SpMat project. So when I was developing this example for use in this book, I first went about fixing R=SS.

We, however, are going to start by making detailed timings of R=RS and R=SR. While we are on the subject of timings, I want to show you how to time code within Mata functions. You time code within functions after making overall timings and discovering a performance problem. You do this in the hope of finding the performance problem and fixing it. The timings for R=SR and R=RS leave room for hope that we might find something, but it will turn out to be in vain. Do not judge the value of performing detailed timings by this one negative result. Detailed timings are often the key to solving performance problems.

18.3   Making detailed timings

Mata provides its own facility for making timings of loops, blocks of code, and even single lines within Mata functions. Those detailed timings can be used to find performance bottlenecks.

Every programmer I know, including me, has been seduced into trying to improve performance by examining the code, finding inefficiencies or inelegancies, and fixing them. The result is that we spend our time improving code where hardly any time was being spent in the first place. The improvements we make are intellectually satisfying but move us no closer to solving performance problems.

The right approach is to find the parts of the code where considerable time is being spent and improve those parts where possible. To make code run faster, you need to make improvements where improvements make a difference.

We will start with R_SxR() and measure, almost line by line, where it is spending its time.

18.3.1   Mata's timer() function

Mata has a suite of timer functions. They are

Mata's timer functions	
Function	Action
timer_clear()	clears all timers
timer_on(#)	start timer #
timer_off(#)	stop timer #
timer()	display report
timer(#)	return value of timer #

  Timers are numbered 1, 2, ..., 100.
  See help mata timer().

You sprinkle calls to timer_on(#) and timer_off(#) around sections of the code. Then, interactively or in do-files, you execute timer_clear(), run the function to be timed—running it repeatedly if it runs too quickly—and finally use timer() to obtain the report.

18.3.2   Make a copy of the code to be timed

To obtain detailed timings, you must modify the source code of the functions to be timed. Do not modify the code in your original file. Copy the original to a new file and make the changes there. The source code for SpMat is in file spmat1.mata. I copied spmat1.mata to a new file that I named code.do:

```
. copy spmat1.mata code.do
```

18.3.3   Make a do-file to run the example to be timed

We want to make detailed timings of R_multiply() when multiplying regular with sparse matrices. We will do that by executing

```
: for (i=1; i<=10; i++) R = R_multiply(A,0, SB,0)
```

where A, B, SA, and SB are the same 1000×1000 banded matrices I used in making the overall timings.

File `timeit.do` will run the timing:

```
─────────────────────────────────────────────── timeit.do ───────
clear all

run code.do
do timingsetup2.do

timer_clear()
for (i=1; i<=10; i++) R = R_multiply(A,0, SB,0)
timer()

end
─────────────────────────────────────────────── timeit.do ───────
```

File `timingsetup2.do` is a copy of `timingsetup.do` with the top lines `clear all` and `run spmat1.mata` removed.

We can run `timeit.do` before we make any changes to the SpMat code, and we will do that to verify that the timing do-file contains no errors:

```
. do timeit
(output omitted )

. ─
```

18.3.4   Add calls to timer_on() and timer_off() to the code

To time code, we add calls to `timer_on(#)` and `timer_off(#)` at the appropriate places in the source code. Where are those appropriate places? We are going to time

```
R = R_multiply(A,0, SB,0)
```

I looked at R_multiply() in `code.do`. It goes through various machinations and calls R_RxS() when given regular and sparse matrices. Let's guess that R_RxS() is the culprit and time it. If we are wrong, we will find out. We have an overall timing of 3.339 seconds. If R_RxS() does not come close to consuming all 3.339 seconds, we can back up and time the other parts of R_multiply().

We will begin by adding calls to **timer_on(1)** and **timer_off(1)** at the top and bottom of **R_RxS()**. Here is the modified **R_RxS()** routine:

```
                                                    ─────── code.do ───────
   .
   .
`Regular´ `Sp´::R_RxS(`boolean´ tB, `Regular´ A)
{
        `Regular´ R
        `RC´       B_j
        `RS´         rA, cA, rB, cB, j

        // R = A*B where B=this (sparse)

        timer_on(1)
        rA = (::rows(A))
        cA = (::cols(A))
        rB = (tB ? cols : rows)
        cB = (tB ? rows : cols)

        if (cA!= rB)  _error(3200)    // conformability error

        R   = J(rA, cB, 0)
        B_j = J(rB,  1, 0)
        for (j=1; j<=cB; j++) {
                set_col(B_j, j, tB)
                R[., j] = A * B_j
        }
        timer_off(1)
        return(R)
}
   .
   .
                                                    ─────── code.do ───────
```

Here is the result from running **timeit.do**:

```
. do timeit
(some output omitted)
: timer_clear()
: for (i=1; i<=10; i++)  R = R_multiply(A,0, SB,0)
: timer()
─────────────────────────────────────────────
timer report
1.        33.5 /          10 =       3.3471
─────────────────────────────────────────────

(some output omitted)
end of do-file

.  ─
```

Look at the timer report produced by `timer()`.

```
timer report
1.      33.5 /        10 =      3.3471
```

The `1.` on the left of the line is the timer number. `33.5` is the total time that timer 1 was on. The `10` before the equals sign reports that the timer was turned on-and-off 10 times. The `3.3471` is the average length of time that the timer ran when it was on.

The report reveals that R_RxS() executes in 3.3471 seconds, which closely matches the overall 3.339 seconds we obtained using `rmsg` earlier. It is in fact longer than 3.339 seconds, which would be impossible were it not the randomness of the timings that we previously discussed. The timings we are producing are accurate to about one-tenth of a second, which we will be accurate enough for our purposes. If we needed more accuracy, we could modify `timeit.do` and run R_multiply() 50 times, or 100.

In any case, we have confirmed that R = R_multiply(A,0, SB,0) is spending its time in R_RxS(), so next we add more timers to the code for R_RxS(). In the modified code below, timer 1 continues to time the entire routine, timer 2 times the routine up to the `for` loop, and timer 3 times the `for` loop itself. Meanwhile, timers 4 and 5 will time the two statements in the body of the loop.

```
———————————————————————————————— code.do ————————
        .
        .
`Regular´ `Sp´::R_RxS(`boolean´ tB, `Regular´ A)
{
        `Regular´ R
        `RC´        B_j
        `RS´        rA, cA, rB, cB, j

        // R = A*B where B=this (sparse)

        timer_on(1)

        timer_on(2)
        rA = (::rows(A))
        cA = (::cols(A))
        rB = (tB ? cols : rows)
        cB = (tB ? rows : cols)

        if (cA!= rB)  _error(3200)     // conformability error

        R    = J(rA, cB, 0)
        B_j = J(rB,   1, 0)
        timer_off(2)

        timer_on(3)
        for (j=1; j<=cB; j++) {
                timer_on(4)
                set_col(B_j, j, tB)
                timer_off(4)

                timer_on(5)
                R[., j] = A * B_j
                timer_off(5)
        }
        timer_off(3)

        timer_off(1)
        return(R)
}
        .
        .
        .
———————————————————————————————— code.do ————————
```

The results of running `timeit.do` are now

```
. do timeit
(some output omitted)
: timer_clear()
: (for i=1; i<=10; i++) R = R_multiply(A,0, SB,0)
```

```
: timer()

timer report
    1.          34 /        10 =       3.4019
    2.        .019 /        10 =        .0019
    3.          34 /        10 =         3.4
    4.        14.7 /     10000 =      .0014743
    5.        19.3 /     10000 =      .0019251
```

(some output omitted)

end of do-file

. _

18.3.5   Analyze timing results

We ran R_multiply() 10 times. The last column of the report reports timers normalized by the number of times each was turned on-and-off. Timers 1, 2, and 3 are normalized by 10 and timers 4 and 5, by 10,000. Timers 4 and 5 are inside a **for** loop of the code being timed, and evidently that **for** loop ran 1,000 times each time we called R_multiply().

Timings will be easier to interpret if we normalize them all by 10 and thus produce averaged results for running r_multiply() once. Those results are

	Total time for running R_RxS() once	
Timer	R_RxS() called by R_multiply(A,0, SB,0)	
1	3.4019 secs.	(overall time)
2	0.0019	(code preceding **for** loop)
3	3.4000	(**for** loop)
2 + 3	3.4019	(equal to timer 1)
4	1.4743	(1st statement, run 1,000 times)
5	1.9251	(2nd statement, run 1,000 times)
4 + 5	3.3994	(nearly equal to timer 3)

Can you interpret the table? The first part of the story is that R_RxS() runs in about 3.4019 seconds and spends nearly all of its time—3.4000 seconds—in the **for** loop. Timers 2 and 3 reveal that.

Thus, we can conclude that whatever the performance bottleneck is, it is inside the loop. The `for` loop is

```
for (j=1; j<=cB; j++) {
        set_col(B_j, j, tB)
        R[., j] = A * B_j
}
```

The first line of the loop sets variable `B_j` equal to the jth column of `B`. The second line sets the jth row of `R` equal to the inner product of `A` and `B_j`. Timers 4 and 5, which time each of the statements, tell the rest of the story:

> 1.4743 seconds—43% of the total—is spent in `set_col()`.
> 1.9251 seconds—57% of the total—is spent in `R[., j] = A * B_j`.

I was relieved and disappointed by these results. I was relieved because I feared that 90 or even 99% of the time might be spent in executing `set_col()`, and that would have been embarrassing to me. It would have meant that I had written a whoppingly inefficient routine. On the other hand, had that been the result, we would have had hope of speeding up the code. We could have planted timers inside `set_col()` and so have discovered `set_col()`'s inefficiency. Instead, we find that `set_col()` executes in about the same time as `R[.,j]=A*B_j` executes, and I know that statement is efficient because it uses Mata's internal operators. The results are ultimately disappointing, however, because I see no way that we can speed up this code without using a new algorithm to multiply sparse matrices.

We will discuss developing new algorithms later. On the timing front, there is still the 16% execution-time difference between `R_RxS()` and `R_SxR()` that perhaps can be eliminated. If I had been smarter, I would have ignored the anomaly and moved on to developing the new algorithms. Even if we find and eliminate the 16% difference— a mere 0.5 seconds—that will still not make `R_SxR()` run fast enough. But instead, I vainly hoped that understanding the anomaly might provide insight I could use to improve the performance of both routines.

So I planted the same timers in `R_SxR()` that we planted in `R_RxS()`. I discovered that the entire 16%, 0.5-second difference was due to the second line of the respective `for` loops, namely,

```
R[., j] = A * B_j      in R_RxS()
R[i, .] = A_i * B      in R_SxR()
```

Remember, this is a 0.5-second difference when each statement is executed 1,000 times. It is a 0.0005-second difference when each statement is executed once.

There is no hope that we can rewrite these lines to execute more quickly. Nonetheless, I was surprised enough that the difference could be traced back to Mata's internal code that I turned the problem over to StataCorp's technical group. You could have done the same by writing to Stata's technical support. They repeated my timings and discovered that on some computers, the first line ran faster; on other computers, the second line ran faster; and on yet other computers, the two lines ran at the same speed. "The difference has to do with cache and other technical features of the chip," they told me.

I put them to needless work. It was doubly needless. Even if they had found a way to eliminate the difference, it would not have solved my problem.

We need new algorithms for R_RxS() and R_SxR(), and we need a new algorithm for R_SxS(), which has a far more serious performance problem. R_SxS() took 21 minutes to execute, not a mere 3.4 seconds. So let's develop better algorithms. It will turn out that the method underlying the better algorithm for R_SxS() can also be applied to R_RxS() and R_SxR().

18.4   Developing better algorithms

We need new algorithms for sparse matrix multiplication. One way to proceed would be to research what others have done. I did not do that because for some problems, there is no literature, and I wanted to show you how to proceed in those cases.

18.4.1   Developing a new idea

We need a new way to multiply sparse matrices. We need an idea. How does one develop ideas? It is a serious question. The obvious answer is to be smart, but I have never found such advice useful. I do not know how to be smarter than I usually am, so what I do instead is get out pencil and paper and work problems by hand.

I multiplied a few sparse matrices. You should try it. If you do, you will discover that you do not blindly follow the formula

$$\mathbf{R} = \mathbf{A} * \mathbf{B} : \ \mathbf{R}_{i,j} = \sum_h \mathbf{A}_{i,h} * \mathbf{B}_{h,j}$$

The first couple of times you multiply matrices by hand, you will follow the formula. Given the matrices

$$\mathbf{A} = \begin{bmatrix} 0 & 0 & 0 & 0 \\ 0 & 2 & 1 & 0 \\ 3 & 0 & 0 & 0 \\ 0 & 0 & 0 & 5 \end{bmatrix}$$

$$\mathbf{B} = \begin{bmatrix} 1 & 0 & 0 & 0 \\ 0 & 0 & 3 & 0 \\ 6 & 0 & 4 & 0 \\ 0 & 0 & 0 & 9 \end{bmatrix}$$

you will follow the formula and calculate

$$(\mathbf{A} * \mathbf{B})_{1,1} = (0 \times 1) + (0 \times 0) + (0 \times 6) + (0 \times 0)$$

$$(\mathbf{A} * \mathbf{B})_{2,1} = (0 \times 1) + (2 \times 0) + (1 \times 6) + (0 \times 0)$$

Later, you will skip terms when one or both elements are 0 and calculate

$$(\mathbf{A} * \mathbf{B})_{1,1} = 0$$

$$(\mathbf{A} * \mathbf{B})_{1,2} = (1 \times 6)$$

You will get amazingly efficient at this. One way to multiply matrices by hand is to write **A** and **B** like this:

$$\mathbf{B} = \begin{bmatrix} 1 & 0 & 0 & 0 \\ 0 & 0 & 3 & 0 \\ 6 & 0 & 4 & 0 \\ 0 & 0 & 0 & 9 \end{bmatrix}$$

$$\mathbf{A} = \begin{bmatrix} 0 & 0 & 0 & 0 \\ 0 & 2 & 1 & 0 \\ 3 & 0 & 0 & 0 \\ 0 & 0 & 0 & 5 \end{bmatrix} \begin{bmatrix} & & & \\ & & & \\ & & & \\ & & & \end{bmatrix}$$

You can now fill in the empty space with the product **A** * **B**. The arrangement of **A** and **B** makes it easy to see how each element of the result is calculated. Each row of **A** and column of **B** points right at the product's element. For instance, the $(1, 1)$ element is produced by matching the first row of **A** with the first column of **B** to produce

$$(\mathbf{A} * \mathbf{B})_{1,1} = (0 \times 1) + (0 \times 0) + (0 \times 6) + (0 \times 0)$$

What you will quickly discover is that you can omit the terms where **A** or **B** is 0. And then you will discover that you can scan across the rows of **A** for the nonzero elements and immediately determine which elements even need to be calculated. The result will be

$$\mathbf{B} = \begin{bmatrix} 1 & 0 & 0 & 0 \\ 0 & 0 & 3 & 0 \\ 6 & 0 & 4 & 0 \\ 0 & 0 & 0 & 9 \end{bmatrix}$$

$$\mathbf{A} = \begin{bmatrix} 0 & 0 & 0 & 0 \\ 0 & 2 & 1 & 0 \\ 3 & 0 & 0 & 0 \\ 0 & 0 & 0 & 5 \end{bmatrix} \begin{bmatrix} 1 \times 6 & & 2 \times 3 + 1 \times 4 & \\ 3 \times 1 & & & \\ & & & 5 \times 9 \end{bmatrix}$$

The elements left blank are 0. Something remarkable just happened. You calculated the product of these 4×4 matrices in just 6 operations instead of the usual 112! At least, that is what happened to me.

I then thought about the 1000×1000 banded matrix that we used in timing the multiplication routines. Multiplying such matrices usually requires just short of two billion calculations. Specifically, it requires $1000 \times 1000 \times (1000 + 999) = 1,999,000,000$ calculations. Given the sparsity of the matrices, however, only 4,996 of the operations—a mere 0.0002% of two billion—produce a nonzero result. Could I reproduce in code the process I used to multiply matrices by hand?

I wrote in a pseudo mathematical/programming notation the approximate process I was following. It is

---

————————————————————————————— idea —————

$\mathbf{R} = \mathbf{A} * \mathbf{B}$, $\mathbf{A}$ and $\mathbf{B}$ sparse:

    Set $\mathbf{R}_{i,j} = 0$ for all i, j.

    For each nonzero value of $\mathbf{A}_{i,h}$ {

        For each nonzero value of $\mathbf{B}_{h,j}$ {

            Update $\mathbf{R}_{i,j} = \mathbf{R}_{i,j} + \mathbf{A}_{i,h} * \mathbf{B}_{h,j}$.

        }

    }

————————————————————————————— idea —————

That is the new idea that I propose we use to rewrite `R_SxS()`.

The question is whether we can write efficient code based on it. The pseudocode contains two "for each nonzero value", but I did not for-each when I multiplied matrices by hand. I scanned for the nonzero values. I scanned across a row for the first for-each and down a column for the second. We humans can scan really quickly. Computers cannot mimic that behavior. Computers access and compare elements one at a time. Computers for-each.

In this case, however, `SpMat` has stored the nonzero elements in an associative array, and the array has at its metaphorical fingertips the subscripts of the nonzero elements. Our code will not need to scan or for-each. Our code can just get the indices corresponding to the nonzero elements from `SpMat`.

18.4.2   Aside

To translate the above idea into code, we will need to use programming tools that may be new to you. It will be better if we discuss them now so that you will be familiar with them later.

To multiply **A** and **B**, we are now planning to work with just the nonzero elements of the matrices. We will obviously need the subscripts (i, j) such that $\mathbf{A}_{i,j} \neq 0$. The first aside below shows you how to obtain them from `AssociativeArray`.

To work with just the nonzero elements of **B**, we will need the j values $jofh(h)$ such that $\mathbf{B}_{h,j} \neq 0$. The problem has some tricky aspects. You need to first understand what $jofh(h)$ is. Consider a matrix in which the nonzero elements in row $h = 1$ are in columns 4 and 9; for row $h = 2$, in columns 8, 9, and 20; and for row $h = 3$, well, there are no nonzero values. One way to summarize this information is in a table:

```
Nonzero elements of sparse matrix B
```

h (row)	$jofh(h)$ (B has nonzero values in columns ...)
1	4, 9
2	8, 9, 20
3	*none*
.	.
.	.

Building $jofh()$ is a programming problem that we will discuss later. Writing the program is not difficult given what `AssociativeArray` can provide to us. What we need to discuss now is how we are going to store $jofh()$. The natural approach would be to store it as a table just as I presented it above. It is an odd sort of table, however, because each row contains a different number of columns. A better way to store tables like this is to make $jofh()$ a pointer vector such that

$$*\texttt{jofh[1]} = 1 \times 2 \text{ row vector equal to } (4, \; 9)$$
$$*\texttt{jofh[2]} = 1 \times 3 \text{ row vector equal to } (8, \; 9, \; 20)$$
$$*\texttt{jofh[3]} = 1 \times 0 \text{ row vector equal to } (\;)$$
$$.$$
$$.$$

This will be the subject of the second aside.

18.4.2.1   Features of associative arrays

Consider the following sparse matrix, which is to say, class `SpMat` scalar `A`:

```
: A.Get(.,.)
       1    2    3    4

  1    0    0    0    0
  2    0    2    1    0
  3    3    0    0    0
  4    0    0    0    5
```

A has four nonzero elements, namely, $(2,2)$, $(2,3)$, $(3,1)$, and $(4,4)$. A stores those nonzero elements in member variable AA, a class `AssociativeArray` scalar. We can think of A.AA as storing

Contents of A.AA	
(Subscripts) Keys	Values
$(2,2)$	2
$(2,3)$	1
$(3,1)$	3
$(4,4)$	5

We can speak about the nonzero elements of A or speak about the values stored in A.AA. Either way, we are speaking about the same elements.

That is important because AA provides functions to tell us about the values it has stored. For instance, function A.AA.N() returns the number of stored values in sparse matrix A. That is an easy and quick way to learn that A has four nonzero values. Even if A were 1000×1000, A.AA.N() would report the number of its nonzero elements just as quickly as it would report 4 for the matrix above. A.AA.keys() does not count the number of nonzero values in A; it merely reports the number of keys it has stored in A.AA.

Function AA.keys() will return the keys. A.AA.keys() returns the number of keys in A.

```
: A.AA.keys()
        1    2
    ┌─────────┐
  1 │  2    2 │
  2 │  4    4 │
  3 │  2    3 │
  4 │  3    1 │
    └─────────┘
```

The four keys listed above are the subscripts of the nonzero values of A. The order in which they are presented—$(2,2)$, $(4,4)$, $(2,3)$, and $(3,1)$—is different from how I presented them, but they are the same subscripts. A.AA stores the keys in a random order, but if we needed them in a particular order, we could use Mata's sort() function. If we needed the keys in row-and-column order, we could code

```
: sort(A.AA.keys(), (1,2))
        1    2
    ┌─────────┐
  1 │  2    2 │
  2 │  2    3 │
  3 │  3    1 │
  4 │  4    4 │
    └─────────┘
```

If we needed them in column-and-row order, we could code

```
: sort(A.AA.keys(), (2,1))
        1    2

1      3    1
2      2    2
3      2    3
4      4    4
```

An important feature of `AA.keys()` is that it is just as fast as `AA.N()` because we are asking for information that `AA` has already recorded. In developing the new idea for multiplying sparse matrices, I scanned the matrices for their nonzero values. `A.AA.keys()` provides the same information for matrix `A`, and it does not even have to scan.

`AA` provides another, related feature. Rather than fetching the entire matrix of subscripts, `AA` provides functions `AA.firstval()` and `AA.nextval()` to step through them one at a time:

```
: for (a=A.AA.firstval(); a; a=A.AA.nextval()) {
>          subscript = A.AA.key()
>          printf("A[%g,%g] = %g\n", subscript[1],
>                    subscript[2], a)
> }
A[2,2] = 2
A[4,4] = 5
A[2,3] = 1
A[3,1] = 3
```

`A.AA.firstval()` and `A.AA.nextval()` return the first and next nonzero value of `A`. When there are no remaining values, the functions return `A.AA`'s not-found value, which SpMat set to 0. Hence, to loop over the nonzero elements, we can code

```
for (a=A.AA.firstval(); a; a=A.AA.nextval()) {
       ...
}
```

Meanwhile, function `AA.key()` will return the key corresponding to the value just returned by `AA.firstval()` or `AA.nextval()`. Those keys are the matrix's subscripts. For matrix `A`, those keys are

```
for (a=A.AA.firstval(); a; a=A.AA.nextval()) {
       subscript = A.AA.key()
       i = subscript[1]
       j = subscript[2]
       ...
}
```

`A.AA.firstval()` and `A.AA.nextval()` step through the values in the order `A.AA` has them stored, which is best thought of as being random. If order matters, we cannot use the `firstval()` and `nextval()` functions. We must instead obtain the subscripts from `A.AA.keys()`, sort them into the desired order, and step through them:

```
subscripts = sort(A.AA.keys(), (1,2))
for (k=1; k<=rows(subscripts); k+++) {
        i = subscript[1]
        j = subscript[2]
        a = A.get(i, j)
        ...
}
```

You can learn more by typing `help mata AssociativeArray`.

18.4.2.2 Advanced use of pointers

I am going to show you how to store a table like

h	j
1	(1)
2	(3)
3	(1, 3)
4	*nothing*

By store, I mean store in such a way that we can quickly access its elements. We want to store the table so that

We can code something like...	and returned will be...	which is a...
$jofh(1)$	(1)	1×1 vector
$jofh(2)$	(3)	1×1 vector
$jofh(3)$	(1, 3)	1×2 vector
$jofh(4)$	( )	1×0 vector
$jofh(1)$`[1]`	1	scalar
.		
.		
$jofh(3)$`[1]`	1	scalar
$jofh(3)$`[2]`	3	scalar
.		
.		

Tables like $jofh$ arise more often in programming than you might guess. They arise when, for each object h, you have a corresponding list of subobjects $jofh(h)$. h could be a bookstore's customer number, and $jofh(h)$ could be the books they have purchased in the last year. We will use h and $jofh(h)$ later in this chapter to be a row of a sparse matrix and the columns in the row containing nonzero values. For instance, we might have a matrix **B** for which

$$jofh(1) = 1$$
$$jofh(2) = 3$$
$$jofh(3) = (1, 3)$$
$$jofh(4) = (\,)$$

That would mean

    in row 1, column 1 is nonzero
    in row 2, column 3 is nonzero
    in row 3, columns 1 and 3 are nonzero
    in row 4, there are no nonzero columns

With $jofh()$ in hand, we will be able to loop across the nonzero elements of sparse matrix B by coding

```
for (h=1; h<=length(jofh); h++) {
        jvalues = jofh(h)
        for (k=1; k<=length(jvalues): k++) {
                j = jvalues[k]
                ... get(B, h, j) ...
        }
}
```

There is no standard word for what $jofh()$ is, although we might call it a matrix with a ragged right edge:

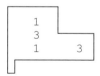

We could say that the matrix is $4 \times (1, 1, 2, 0)$.

I would be more tempted to call $jofh()$ a list of lists because I think of its contents—(1), (3), (1,3), and ( )—as lists of column values. Whatever we call this strange object, we can create it using pointers.

Until now, we have used pointers as synonyms for existing variables. For instance, you are already familiar with constructs such as

```
: p = &myscalar
: *p = 2
```

The first statement puts an address in **p** (the address of variable **myscalar**), and the second statement stores 2 at that address.

Here is another construct that results in ***p==2**, but **p** will not contain the address of **myscalar** or of any existing variable:

```
: p = & (2)
```

The **(2)** in the above could just as well be any parenthesized expression, such as

```
: p = & (3-1)
```

Either of these statements results in the expression inside the parentheses being evaluated to produce 2. The **&** in front of the parentheses extracts the address of where the 2 is stored, and that address is put in **p**.

All three of the above examples result in ***p==2**. The last two examples, however, are different in that the address stored in **p** corresponds to no existing variable.

We can use this approach to create $jofh()$. We can create a 4×1 pointer vector **jofh** by typing

```
: jofh = J(4, 1, NULL)
```

At this point, **jofh** is a column vector with each element set to **NULL**. We can reset each of its elements to the address of a vector containing the appropriate values:

```
: jofh[1] = & (1)              // address of vector = (1)
: jofh[2] = & (3)              // address of vector = (3)
: jofh[3] = & (1, 3)           // address of vector = (1, 3)
: jofh[4] = & (J(1, 0, .))     // address of 1 x 0 vector
```

If we were using **jofh** in a program, its declaration would be

```
pointer(real vector) vector   jofh
```

Here is how we could access the values stored in *jofh:

We can type...	and returned will be...	which is a...
*jofh[1]	(1)	1×1 vector
*jofh[2]	(3)	1×1 vector
*jofh[3]	(1, 3)	1×3 vector
*jofh[4]	( )	1×0 vector
(*jofh[1])[1]	1	scalar
.		
.		
(*jofh[3])[1]	1	scalar
(*jofh[3])[2]	3	scalar
.		
.		

For instance, code for looping across the nonzero values of SpMat matrix B could read

```
pointer(real vector) vector    jofh
pointer(real vector) scalar    jvalues
real scalar                    h, el, j
for (h=1; h<=length(jofh); h++) {
        jvalues = jofh[h]
        for (el=1; el<=length(*jvalues): el++) {
                j = *jvalues[el]
                ... get(B, h, j) ...
        }
}
```

In describing jofh, I have imagined that *jofh[h] is a 1×0 vector when there are no
j values. Instead of filling in empty elements with the address of null vectors, we could
instead simply leave the NULL pointer value in place.

```
: jofh = J(4, 1, NULL)

: jofh[1] = & (1)
: jofh[2] = & (3)
: jofh[3] = & (1, 3)
: //                            jofh[4] = NULL  already
```

If we did that, we would need to add an `if` statement to the code for looping across the nonzero elements of B:

```
for (h=1; h<=length(jofh); h++) {
        jvalues = jofh[h]
        if (jvalues) for (el=1; el<=length(*jvalues): el++) {
                j = *jvalues[el]
                ... get(B, h, j) ...
        }
}
```

In the above code, the `for` loop will be executed only when `jvalues` is not equal to NULL. As far as conditional statements are concerned, pointers equal to NULL are treated just like real values equal to 0. Both are treated as false.

In appendix D, I show another advanced, nonsynonym use of pointers. I use pointers to create three-dimensional arrays.

18.5   Converting the new idea into code sketches

We are about to convert the idea we have sketched for multiplying sparse matrices into code sketches. Before we do that, I want to emphasize the three new tricks we just learned in the previous section.

Trick 1. We can loop through all the nonzero elements of SpMat matrix A, albeit in random order, by coding

```
for (a=A.AA.firstval(); a; a=A.AA.nextval()) {
        subscript = A.AA.key()
        i = subscript[1]
        j = subscript[2]
        ...
}
```

In the above code, i and j are the subscripts of each nonzero element, and a is its value.

Trick 2. When order matters, we can loop through in row-column order by coding

```
subscripts = sort(A.AA.keys(), (1,2))
for (k=1; k<=rows(subscripts); k++) {
        i = subscripts[k, 1]
        j = subscripts[k, 2]
        a = A.get(i, j)
        ...
}
```

Trick 3. Unrelatedly, we can create a pointer object `jofh` such that `*jofh[h]` is a vector. The vectors can be of different lengths.

We will need all three tricks to translate the idea into code.

18.5.0.3   Converting the idea into a sketch of R_SxS()

The idea we need to translate is

── idea ─────────

$\mathbf{R} = \mathbf{A} * \mathbf{B}$, $\mathbf{A}$ and $\mathbf{B}$ sparse:

Set $\mathbf{R}_{i,j} = 0$ for all i, j.

For each nonzero value of $\mathbf{A}_{i,h}$ {

For each nonzero value of $\mathbf{B}_{h,j}$ {

Update $\mathbf{R}_{i,j} = \mathbf{R}_{i,j} + \mathbf{A}_{i,h} * \mathbf{B}_{h,j}$.

}

}

── idea ─────────

The translated idea will be the sketch of the new `R_SxS()` function. Translating the first line—set $\mathbf{R}_{i,j} = 0$ for all i, j—is easy enough. As we fill in the code, the translated lines will appear as `typewriter` text, while the untranslated lines will continue to appear in normal typeface:

────────────────────── partially translated `R_SxS()` ─────────

```
class SpMat scalar  A, B       // matrices to be multiplied
class SpMat scalar  R          // result: R = A*B
R = J(A.rows(), B.cols(), 0)
```

For each nonzero value of $\mathbf{A}_{i,h}$ {

For each nonzero value of $\mathbf{B}_{h,j}$ {

Update $\mathbf{R}_{i,j} = \mathbf{R}_{i,j} + \mathbf{A}_{i,h} * \mathbf{B}_{h,j}$.

}

}

────────────────────── partially translated `R_SxS()` ─────────

The next line to translate is

For each nonzero value of $\mathbf{A}_{i,h}$ {

We can translate the line using trick 1 because the order in which we step through the elements of A does not matter:

——————————————————————— partially translated R_SxS() ————————

```
class SpMat scalar   A, B
class SpMat scalar   R

real scalar          a
real scalar          i, h

R = J(A.rows(), B.cols(), 0)
for (a=A.AA.firstval(); a; a = A.AA.nextval()) {
        subscript = A.AA.key()
        i          = subscript[1]
        h          = subscript[2]
```

For each nonzero value of $\mathbf{B}_{h,j}$ {

Update $\mathbf{R}_{i,j} = \mathbf{R}_{i,j} + \mathbf{A}_{i,h} * \mathbf{B}_{h,j}$.

}

```
}
```

——————————————————————— partially translated R_SxS() ————————

The remaining for-each statement to be translated is an instruction to loop over the nonzero elements of row h of B.

For each nonzero value of $\mathbf{B}_{h,j}$ {

firstval() and nextval() cannot do that, so let's assume a trick-3 object jofh such that *jofh[h] is the vector of j values such that $\mathbf{B}_{h,j}$ is nonzero. Then we merely need to loop over the values of *jofh[h]. (We now use square brackets and not parentheses because jofh is a vector, not a function.)

Let's rename jofh to have a more meaningful name, cols_of_row. cols_of_row will be pointer(real vector) vector.

Let's also assume the existence of a function `B.cols_of_row()`, which we can call to fill in the variable `cols_of_row()`. In Mata, functions and variables can have the same names, so do not confuse them in the translated code, which is

──────────────────────────────── partially translated `R_SxS()` ────────

```
class SpMat scalar          A, B
class SpMat scalar          R

real scalar                 a
real scalar                 i, h, j, k
pointer(real vector) vector cols_of_row
pointer(real vector) scalar p
R           = J(A.rows(), B.cols(), 0)
cols_of_row = B.cols_of_row()
for (a=A.AA.firstval(); a; a = A.AA.nextval()) {
        subscript = A.AA.key()
        i           = subscript[1]
        h           = subscript[2]
        p           = cols_of_row[h]
        if (p) for (k=1; k<=length(*p); k++) {
                j = (*p)[k]
```

$$\text{Update } \mathbf{R}_{i,j} = \mathbf{R}_{i,j} + \mathbf{A}_{i,h} * \mathbf{B}_{h,j}.$$

```
        }
}
```

──────────────────────────────── partially translated `R_SxS()` ────────

We have just one line left to translate:

$$\text{Update } \mathbf{R}_{i,j} = \mathbf{R}_{i,j} + \mathbf{A}_{i,h} * \mathbf{B}_{h,j}.$$

The line almost translates itself:

```
——————————————————————————————— completed sketch of R_SxS() —————————

class SpMat scalar          A, B
class SpMat scalar          R

real scalar                 a
real scalar                 i, h, j, k

pointer(real vector) vector   cols_of_row
pointer(real vector) scalar   p

R           = J(A.rows(), B.cols(), 0)
cols_of_row = B.cols_of_row()
for (a=A.AA.firstval(); a; a = A.AA.nextval()) {
        subscript = A.AA.key()
        i           = subscript[1]
        h           = subscript[2]
        p           = cols_of_row[h]
        if (p) for (k=1; k<=length(*p); k++) {
                j = (*p)[k]
                R[i, j] = R[i, j] + a*B.get(h, j)
        }
}

——————————————————————————————— completed sketch of R_SxS() —————————
```

We now have a sketch of a new R_SxS().

18.5.0.4 Sketching subroutine cols_of_row()

The new R_SxS() function calls new subroutine B.cols_of_row(). B.cols_of_row() returns a pointer(real vector) vector. That is a mouthful, but the idea is simple enough. For each row i of B, *cols_of_row[i] will be a vector containing the values of j such that $B_{i,j}$ is not equal to 0.

I say matrix B because that is what we called it when we used cols_of_row() in the sketch of R_SxS(). I am going to call the matrix A now that we are sketching the cols_of_row() function. The sketch begins

```
——————————————————————————————————— sketch of cols_of_row() —————————
pointer(real vector) vector SpMat::cols_of_row()
{
        pointer(real vector) vector  p
        .
        .
        return(p)
}
——————————————————————————————————— sketch of cols_of_row() —————————
```

Note that I am calling the value to be returned **p**. To understand what we need to put into **p**, let's consider an example. Here is a sparse matrix:

```
: A.Get(.,.)
       1    2    3    4

  1    0    0    0    0
  2    0    2    1    0
  3    3    0    0    0
  4    0    0    0    5
```

For this matrix, **p** will have four elements because the matrix has four rows. The contents of **p** will be

```
 p[1]  = NULL       // row 1 of A has no nonzero elements
*p[2]  = (2, 3)     // row 2 of A has nonzero elements in columns 2 and 3
*p[3]  = (1)        // row 3 of A has nonzero elements in column 1
*p[4]  = (4)        // row 4 of A has nonzero elements in column 4
```

We will derive the contents of **p** from the sorted subscripts of **A**'s nonzero elements:

```
: sort(A.AA.keys(), (1,2))
       1    2

  1    2    2
  2    2    3
  3    3    1
  4    4    4
```

The first row of this matrix says that $A_{2,2}$ is nonzero; the second row, that $A_{2,3}$ is nonzero; and so on. We can read the results to be stored in **p[]** directly from it:

sort(A.AA.keys(), (1,2))		Result
Row 1 is (2, 2) and Row 2 is (2, 3)	means	*p[2] = (2, 3)
Row 3 is (3, 1)	means	*p[3] = (1)
Row 4 is (4, 4)	means	*p[4] = (4)

Here is a sketch of the code to create p:

```
────────────────────────────────── sketch of cols_of_row() ──────────
pointer(real vector) vector SpMat::cols_of_row()
{
        pointer(real vector) vector  p
        real matrix                  subscripts
        real scalar                  r, i, i1, ii

        p = J(A.rows(), 1, NULL)
        subscripts = sort(A.AA.key(), (1,2))

        for (i=1; i<=rows(subscripts); i=i1) {
                r = subscripts[i, 1]

                // ---------------------------------------------
                // subscripts for row r start at subscripts[i,]
                // they continue until subscripts[i1-1,]
                // Find i1:

                for (i1=i+1; i1<=rows(subscripts); i1++) {
                        if (subscripts[i1, 1] != r) break
                }

                // ---------------------------------------------
                // Make *p[r] a 1 x i1-i real vector:

                p[r] = &( J(1, i1-i, .) )

                // ---------------------------------------------
                // Fill in *p[r]:

                for (ii=i; ii<i1; ii++) {
                        (*p[r])[++k] = subscripts[ii, 2]
                }
        }

        return(p)
}
────────────────────────────────── sketch of cols_of_row() ──────────
```

I have highlighted the comments and a little of the code in lieu of providing a detailed explanation. The outer i loop loops across the sorted table of subscripts. Inside the loop, the code finds i1 such that subscripts i to i1-1 all refer to row r. The code creates and fills in *p[r] and then starts again with i=i1.

18.5.1 Converting sketches into completed code

We now have completed sketches of R_SxS() and cols_of_row(). The plan is to produce final code from them and measure the performance of the new code. If it is fast enough, we can then apply the same idea to improve the performance of R_SxR() and R_RxS().

File spmat1.mata contains the current, certified SpMat code. I copied it to new file spmat2.mata,

```
. copy spmat1.mata spmat2.mata
```

and I modified its header to state that this file contains version 2.0.0 code:

```
——————————————————————— excerpt from spmat2.mata ———————
*! version 2.0.0

version 15
set matastrict on
    .
    .
——————————————————————— excerpt from spmat2.mata ———————
```

I added the declaration of `cols_of_row()` to the class declaration and added new macroed type 'RV', meaning real vector:

```
——————————————————————— excerpt from spmat2.mata ———————
    .
    .
local RV        real vector
    .
    .
class `Sp´
{
        public:
            .
            .
        private:
            .
            .
        pointer(`RV´) vector    cols_of_row()
            .
            .
}
    .
    .
——————————————————————— excerpt from spmat2.mata ———————
```

The sketch we made of `R_SxS()` is almost completed code. It lacked certain details, such as code to ensure that the matrices to be multiplied are conformable, and it lacked a few declarations. I completed the code and put it in **spmat2.mata**. Starting at the line that reads "// ------ multiply ---", the code matches the sketch except that I added `q=p[h]` because I thought using `q` instead of `p[h]` in the remaining code made it more readable.

```
                                                        ————————— spmat2.mata ————
359:  `Regular´ `Sp´::R_SxS(`boolean´ tA, `Sparse´ B, `boolean´ tB)
360:  {
361:          `Regular´  R                //             R = A*B
362:          `RS´       v                //             v = A[i,h]
363:          `RR´       subscript        // subscript = (i,h)
364:          `RS´       r, c, d          // subscript bounds
365:          `RS´       i, j, h          // subscript indices
366:          `RS´       k, n
367:
368:          pointer(`RV´) vector    p
369:          pointer(`RV´) scalar    q
370:
371:          // ------------------------------ code limitation ---
372:          if (tA | tB) {
373:                  _error(9, "!! tA & tB not yet implemented")
374:                  // NotReached
375:          }
376:
377:          // -------------------------- check conformability ---
378:          /*
379:              R_SxS() calculates R=A*B for A, B sparse.
380:              Notation is
381:                  matrix        dimension    indices
382:                  ------------------------------------
383:                    R             r x c        i, j
384:                    A (this)      r x d        i, h
385:                    B             d x c        h, j
386:                  ------------------------------------
387:          */
388:
389:          r = this.rows ; d = this.cols        // A:  r x d
390:          c = B.rows                           // B:  d x c
391:                                                    // R:  r x c
392:          if (d != B.cols) {
393:                  _error(3200)                 // conformability error
394:                  // NotReached
395:          }
396:
397:          // -------------------------------------- multiply ---
398:          p = B.cols_of_row(tB)
399:          R = J(r, c, 0)
400:          for (v=AA.firstval(); v; v=AA.nextval()) {
401:                  subscript = AA.key()
402:                  i         = subscript[1]
403:                  h         = subscript[2]
404:                  if ( (q=p[h]) ) {
405:                          n = length(*q)
406:                          for (k=1; k<=n; k++) {
407:                                  j = (*q)[k]
408:                                  R[i, j] = R[i, j] + v*B.get(h, j)
409:                          }
410:                  }
411:          }
412:          return(R)
413:  }
                                                        ————————— spmat2.mata ————
```

When I did all the above, I was in a rush to obtain the new timings. As a result, I decided to omit adding code to handle R_SxS()'s arguments tA and tB. Those arguments specify that the matrices be treated as if they were transposed. Instead of writing code to do that, I wrote

```
if (tA | tB) {
        _error(9, "!! tA & tB not yet implemented")
        // NotReached
}
```

This code aborts execution if tA or tB is true. There are two things to explain about this code: the use of !! and the use of // NotReached.

18.5.1.1 Double-bang comments and messages

When I write code that has shortcomings that I must fix before it can be considered complete, I include comments or error messages, and I prefix the text of the comments and messages with !!. When I am in the midst of development, code literally can be sprinkled with double-bang comments and error messages, such as,

```
// !! is this right?
// !! I suspect there is better formula
// !! The following loop is inefficient but works.
// !! include certification test for n=k=0
// _error("!! tA & tB not yet implemented")
```

When something occurs to me that I need to do later, I write a double-bang comment or error. I put them in files containing Mata code, files containing certification scripts, files containing documentation, and even files containing design documents. No file is immune.

When I am in the mood to clean up details, or a deadline looms, I search my files for !! and fix them, or fix some of them and leave others for later. There is no rush, but no file is complete and no project is finished until all double-bangs are found, dealt with, and removed.

I use double-bangs as reminders for little things and big things. Double-bangs are my to-do lists. When writing serious code, the excuse "I forgot" is unacceptable.

18.5.1.2   // NotReached comments

In the code snippet I just showed you,

```
if (tA | tB) {
        _error(9, "!! tA & tB not yet implemented")
        // NotReached
}
```

the `// NotReached` is just a comment. I include this comment when it is not otherwise
visually obvious that flow of control has already ended. For instance, function `_error()`
displays the message and aborts execution. The `// NotReached` makes it explicit that
`_error()` returned or aborted.

Had the line read

```
if (tA | tB) return
```

or

```
if (tA | tB) exit(198)
```

I would usually not bother with a `// NotReached` comment. The statements themselves
make it clear that execution does not proceed beyond this point. On the other hand,
it is sometimes worth emphasizing with a `// NotReached` when the statements are
unexpected. However, I will not put `// NotReached` comments even on surprising
statements when there would be too many of them in a small space, or if they would
interfere with readability of the code. The purpose of `// NotReached` is emphasis to
improve readability of the code, not to diminish it.

18.5.1.3   Back to converting sketches

We have one more routine to convert from sketch to final code before we can evaluate the
performance of R_SxS(), namely, `cols_of_row()`. The final code is nearly identical to
the sketch. The biggest difference is the unexpected inclusion of an extra argument `tA`.
When I placed the double-banged not-yet-implemented message in R_SxS(), I realized
that `cols_of_row()` would also need a virtual transposition capability, so I added the
argument and another double-banged not-yet-implemented message.

The final-but-for-double-bangs code is

```
──────────────────────────────────────────────── spmat2.mata ──────────
416:  pointer(`RR´) vector `Sp´::cols_of_row(`boolean´ tA)
417:  {
418:          pointer(`RV´) vector p           // to be returned
419:          pointer(`RV´) scalar q           // q = p[r] for some r
420:          `RM´                    subscripts // (row,col) values
421:          `RS´                    r          // row of matrix
422:          `RS´                    i, i1, ii // subscripts for subscripts[]
423:          `RS´                    k          // subscript  for q[]
424:
425:          if (tA) _error(9, "!!tA not yet implemented")
426:
427:          p           = J(this.rows, 1, NULL)
428:          subscripts = AA.keys()
429:          if (::rows(subscripts) == 0) return(p)
430:
431:          _sort(subscripts, (1,2))
432:
433:          pragma unset i1
434:          for (i=1; i<=(::rows(subscripts)); i=i1) {
435:                  r  = subscripts[i, 1]
436:
437:                  // -------------------------------------------------
438:                  // Indices for row r start at subscripts[   i, .]
439:                  // They continue until       subscripts[i1-1, .]
440:                  // Find i1:
441:
442:                  for (i1=i+1; i1<=(::rows(subscripts)); i1++) {
443:                          if (subscripts[i1, 1] != r) break
444:                  }
445:
446:                  // -------------------------------------------------
447:                  // Make a vector containing i1-i elements and
448:                  // store the column subscripts in it:
449:
450:                  q = p[r] = &(J(1, i1-i, .))
451:                  k = 0
452:                  for (ii=i; ii<i1; ii++) (*q)[++k] = subscripts[ii,2]
453:          }
454:          return(p)
455:  }
──────────────────────────────────────────────── spmat2.mata ──────────
```

18.5.2 Measuring performance

I reran the timing of R_SxS(), the timing we called T5 back at the beginning of this chapter. Here is what I found:

S_RxR()	Time to multiply two 1000 × 1000 sparse matrices
Before (spmat1.mata)	1,266.68 seconds
After (spmat2.mata)	0.0316 seconds

You might take a moment to appreciate this result. One does not often see speed improvements of four million percent. We need to do more with different sparse matrices to confirm these results, but I am confident that this timing is not a fluke because the run time of the new logic is a function of the number of nonzero elements stored in the matrices instead of the number of rows and columns. The more sparse the matrix, the faster R_SxS() will run.

The SpMat project is now back on track.

Sometimes algorithms matter.

18.6 Cleaning up

Cleaning up means doing what is necessary so that code can pass certification. Before running the timing reported above, I did a little bit that I did not tell you about to verify that the new code produces correct answers. I verified that it produced correct results when multiplying the 1000 × 1000 banded matrices used in the timing.

I had not yet run the certification script because I could not run it. R_SxS() and cols_of_row() do not yet handle virtual transposition. We need to substitute code for double-bangs in R_SxS() and cols_of_row().

The current, double-banged code is in **spmat2.mata**. I copied **spmat2.mata** to new file **spmat3.mata**,

```
. copy spmat2.mata spmat3.mata
```

and modified the new file to have an even newer internal version number:

```
──────────────────────────── excerpt from spmat3.mata ────────────
*! version 3.0.0
version 15
set matastrict on
 .
 .
──────────────────────────── excerpt from spmat3.mata ────────────
```

The plan now is to modify the code in `spmat3.mata` to handle virtual transposition, and if something goes wrong, we will still have the `spmat2.mata` code so that we can start again.

18.6.1 Finishing R_SxS() and cols_of_row()

I added the code to `R_SxS()` and `cols_of_row()` to handle virtual transposition. The additions were not obvious, yet you could have made them. The updated code is

```
                                                     ─── spmat3.mata ────

346:    `Regular´ `Sp´::R_SxS(`boolean´ tA, `Sparse´ B, `boolean´ tB)
347:    {
348:            `Regular´  R                //           R = A*B
349:            `RS´       v                //           v = A[i,h]
350:            `RR´       subscript        // subscript = (i,h)
351:            `RS´       r, c, d          // subscript bounds
352:            `RS´       i, j, h          // subscript indices
353:            `RS´       k, n
354:            `RS´       row, col         // index of row & col in subscript[]
355:
356:        pointer(`RV´) vector    p
357:        pointer(`RV´) scalar    q
358:
359:
360:            /*
361:               R_SxS() calculates R=A*B for A, B sparse.
362:               Notation is
363:                    matrix        dimension    indices
364:                    -------------------------------------
365:                    A (this)      r x d          i, h
366:                    B             d x c          h, j
367:                    R             r x c          i, j
368:                    -------------------------------------
369:            */
370:
371:            if (tA) {
372:                    r   = this.cols
373:                    d   = this.rows
374:                    row = 2          // subscript[2] is row
375:                    col = 1          // subscript[1] is col
376:            }
377:            else {
378:                    r = this.rows
379:                    d = this.cols
380:                    row = 1          // subscript[1] is row
381:                    col = 2          // subscript[2] is col
382:            }
383:
384:            if (tB) {
385:                    if (d != B.cols) _error(3200) // conformability
386:                    c = B.rows
387:            }
388:            else {
389:                if (d != B.rows) _error(3200) // conformability
390:                    c = B.cols
391:            }
```

```
392:
393:          p = B.cols_of_row(tB)
394:
395:          R = J(r, c, 0)
396:
397:
398:          /*
399:              Note:  The only difference between the two
400:              branches below is
401:                        R[i,j] = R[i,j]+ v*B.get(j,h)   ( tB)
402:                        R[i,j] = R[i,j]+ v*B.get(h,j)   (!tB)
403:              There are separate branches so that this deeply nested
404:              statement executes as quickly as possible.
405:          */
406:
407:          if (tB) {
408:                  for (v=AA.firstval(); v; v=AA.nextval()) {
409:                          subscript = AA.key()
410:                          i    = subscript[row]
411:                          h    = subscript[col]
412:                          if ( (q=p[h]) ) {
413:                                  n = length(*q)
414:                                  for (k=1; k<=n; k++) {
415:                                          j = (*q)[k]
416:                                          R[i,j] = R[i,j] + v*B.get(j,h)
417:                                  }
418:                          }
419:                  }
420:          }
421:          else {
422:                  for (v=AA.firstval(); v; v=AA.nextval()) {
423:                          subscript = AA.key()
424:                          i    = subscript[row]
425:                          h    = subscript[col]
426:                          if ( (q=p[h]) ) {
427:                                  n = length(*q)
428:                                  for (k=1; k<=n; k++) {
429:                                          j = (*q)[k]
430:                                          R[i,j] = R[i,j] + v*B.get(h,j)
431:                                  }
432:                          }
433:                  }
434:          }
435:          return(R)
436:  }
```

—————————————————————————————— spmat3.mata ——————

```
──────────────────────────────────────────────── spmat3.mata ─────────
445: pointer(`RR´) vector `Sp´::cols_of_row(`boolean´ tA)
446: {
447:         pointer(`RV´) vector p          // to be returned
448:         pointer(`RV´) scalar q          // q = p[r] for some r
449:         `RM´                   subscripts // (row,col) values
450:         `RS´                   r          // row of matrix
451:         `RS´                   i, i1, ii  // subscripts for subscripts[]
452:         `RS´                   k          // subscript  for q[]
453:         `RS´                   row        // 1 or, if tA, 2
454:         `RS´                   col        // 2 or, if tA, 1
455:         `RS´                   rows       // "rows" of this
456:
457:         if (tA) {
458:                 row  = 2
459:                 col  = 1
460:                 rows = this.cols
461:         }
462:         else {
463:                 row  = 1
464:                 col  = 2
465:                 rows = this.rows
466:         }
467:
468:         p           = J(rows, 1, NULL)
469:         subscripts = AA.keys()
470:         if (::rows(subscripts) == 0) return(p)
471:
472:         _sort(subscripts, (row,col))
473:
474:         pragma unset i1
475:         for (i=1; i<=(::rows(subscripts)); i=i1) {
476:                 r  = subscripts[i, row]
477:
478:                 // Indices for "row" r start at subscripts[   i,]
479:                 // They continue until        subscripts[i1-1,]
480:                 // Find i1:
481:
482:                 for (i1=i+1; i1<=(::rows(subscripts)); i1++) {
483:                         if (subscripts[i1, row] != r) break
484:                 }
485:
486:                 // Make a vector containing i1-i elements and
487:                 // store the column subscripts in it.
488:
489:                 q = p[r] = &(J(1, i1-i, .))
490:                 k = 0
491:                 for (ii=i; ii<i1; ii++) (*q)[++k] = subscripts[ii,col]
492:         }
493:         return(p)
494: }
──────────────────────────────────────────────── spmat3.mata ─────────
```

After I had modified `cols_of_row()` to add transposition, I realized that `cols_of_row()` with transposition amounted to performing rows-of-column. I decided to add a `rows_of_col()` function in case I, or some other programmer, needed it later:

```
──────────────────────────────────── rows_of_col_ptr() ─────────
pointer(`RR´) vector `Sp´::rows_of_col(`boolean´ tA)
{
        return(cols_of_row_ptr(!tA))
}
──────────────────────────────────── rows_of_col_ptr() ─────────
```

Functions `cols_of_row()` and `rows_of_col()` are now complicated enough that I wanted to see for myself that they returned correct results for some simple examples. I was pretty sure that `cols_of_row()` with $tA = 0$ would work. I really wanted to see it work with $tA = 1$. So I temporarily modified the code to make the functions public, and they did work. I realized that I might someday need to modify the functions. Perhaps I would need to make them run faster. If so, it would save me time if I included the tests that I performed interactively in the certification script. So I left the functions as public. The declaration of **SpMat** now read

```
──────────────────────────────── excerpt from spmat3.mata ─────────
    .
    .
    .
class `Sp´
{
    public:
        .
        .
        .
    private:
        .
        .
        .
    public /*DND*/:
        pointer(`RV´) vector   cols_of_row()
        pointer(`RV´) vector   rows_of_col()
}
    .
    .
    .
──────────────────────────────── excerpt from spmat3.mata ─────────
```

I marked the functions as DND to remind myself that these functions need not be documented.

You can see a complete listing of **spmat3.mata** by typing in Stata

```
. view ~/matabook/spmat3.mata
```

If you have not yet downloaded the files associated with this book, see section 1.4.

18.6.2   Running certification

I ran certification. I copied **test_spmat1.do** to new file **test_spmat3.do**,

```
. copy test_spmat1.do test_spmat3.do
```

I modified the new file to load the code from **spmat3.mata**:

```
──────────────────────────── excerpt from test_spmat3.do ────────────
// version number intentionally omitted

clear all
run spmat3.mata
.
.
.
──────────────────────────── excerpt from test_spmat3.do ────────────
```

The do-file ran without error. I then added the direct tests of **cols_of_row()** and **rows_of_col()** to the end of the file and ran the do-file again. Finally, I added a few more tests of multiplying **SpMat** matrices. This time, I created matrices that were not sparse but had been stored as **SpMats** anyway. That is, I created

```
A = runiform(5,4)
B = runiform(4,5)
```

and then made **SpMat** matrices **SA** and **SB** from them. Sparse or not, multiplying **SA*SB** should produce the same result as **A*B**:

```
. do test_spmat3.do
  (output omitted )
:
: R = R_multiply(SA,0, SB,0)
: assert(R==A*B)
        assert():  3498  assertion is false
          <istmt>:     -  function returned error
(3498);
```

The code failed the test! I wondered how different the matrices were. I temporarily modified the test script to show how similar **R** and **A*B** were and reran it:

```
. do test_spmat3.do
  (output omitted )
:
: R = R_multiply(SA,0, SB,0)
: mreldif(R, A*B)                              // <-- new
  1.02810e-16
: assert(R==A*B)
        assert():  3498  assertion is false
          <istmt>:     -  function returned error
(3498);
```

Mata function `mreldif(R, A*B)` returns the maximum relative difference of the elements of `R` and `A*B`. A maximum relative difference of 10^{-16} means that when comparing the two results, the largest difference was in the sixteenth digit. The first fifteen digits always match. Fifteen digits of accuracy is right at the limit of double precision. Matrices `R` and `A*B` may not be exactly equal, but they are equal.

Matrix multiplication involves lots of additions and, as we discussed in section 13.4.1, just changing the order in which the additions are performed can change numerical results. The order is different in this case because of how `R_SxS()` performs matrix multiplication. It steps across the nonzero elements using the `firstval()` and `nextval()` functions. The order in which `firstval()` and `nextval()` return those nonzero values determines the order in which the additions are performed.

I changed the certification script code to read

```
. do test_spmat3.do
  (output omitted)

:
: R = R_multiply(SA,0, SB,0)
: assert(mreldif(R, A*B) < 1e-15)
  1.02810e-16
  (output omitted)

. _
```

`spmat3.mata` passed certification.

You can see the full listing of **test_spmat3.do** by typing in Stata

```
. view ~/matabook/test_spmat3.do
```

18.7   Continuing development

We can now continue development, although I hope you will excuse me if I drop out. You may want to join me. The purpose of this and the previous chapter has been to show you how to approach a large development project and how to drive it to completion. I think we have accomplished that. In fact, I would judge that we are halfway to completing the project.

I am sure that interesting issues would arise if we continued. Some of the issues might even require that we be smarter and cleverer than we usually are. In those cases, we would get out pencil and paper and work examples by hand in hopes that the exercise leads to inspiration. It usually does, in my experience.

I will give you a list of what remains to be done, not because I expect you to complete this project, but as part of the training for how to drive large projects to completion. The next step is to make precisely that list. Here it is.

1. Make R_SxR() and R_RxS() efficient using the approach we used together to make R_SxS() efficient. Check performance and certify code.

2. Make S_*x*() efficient. This is a moderate-sized project. I would first work on S_SxS(), update certification, and then work on the remaining routines. You will be able to borrow ideas used in R_*x*(), but details will need to vary if code is to run quickly. You do not want to code statements in the innermost multiplication loop such as

   ```
   R.Put( R.get(i,j) + v*B.get(h,j)
   ```

   I say that, but I would still try it. It might work well enough. I have shown you two second-rate approaches in this book and neither worked out. They were not random choices. I wanted to show you how easy it is to switch algorithms in well-written code, and to show you that developing code is easier using second-rate algorithms. What I have not shown you is that sometimes the second-rate algorithms work well enough. This might be one of those cases.

   If you find that the second-rate algorithms do not perform adequately, the fix is to code

   ```
   summarize = sum + v*B.get(h,j)
   ```

   and to R.put(sum, i, j) after the summation is complete. To do this, you will need to reorder the for loops to loop first through i, then through j, and finally through h.

3. Make the S_*p*() matrix-addition functions efficient. You will need to develop an idea along the lines of the idea we developed for sparse matrix multiplication. The idea for addition will be simpler than the idea was for multiplication, but the resulting code will be approximately as complicated to implement. I recommend that you use AA.keys() to get the nonzero elements of both matrices and then form lists of those elements that can be copied from A into the result, those that can be copied from B, and those that require addition of the corresponding nonzero elements of A and B.

There are three more items on the list, but they will each be quick and easy:

4. Write S_*m*() functions to subtract matrices. Start with the S_*p*() routines and modify them.

5. Add a transposition function Aprime = A.transpose().

6. Add a negation function Aneg = A.negate().

As I said, we are already halfway done.

19 The Mata Reference Manual

Type `help mata` in Stata or Mata, and you will see

Section	Description
[M-1]	Introduction and advice
[M-2]	Language definition
[M-3]	Commands for controlling Mata
[M-4]	Categorical guide to Mata functions
[M-5]	Alphabetical index to Mata functions
[M-6]	Mata glossary of common terms

You are looking at the online version of the *Mata Reference Manual*. I started this book by making harsh comments about the manual. I compared it with a parts manual for a car. Now that you have seen how those parts can be fit together, I suggest you give Mata's parts manual another chance. The manual is comprehensive if concise. This book was the opposite.

Type `help mata` and click on [M-1], [M-2], and [M-3].

Read sections [M-1] through [M-3] in their entirety. They are not only a good review of what we covered but they also include details that the book ignored or glossed over. For instance, [M-1] includes a discussion of absolute and relative tolerances, a topic that this book ignored, and the manual has a more thorough discussion of permutation vectors. If pointers still confuse you, read the discussion and examples in [M-2]. And we did not even discuss in this book the material in [M-3]. That material is not of great importance to serious programmers, but there is one feature that I use daily: `mata which`. I will leave it to you to learn about it.

Click on [**M-4**] for the built-in functions part of Mata's parts manual. Every built-in function is categorized and described. When you click on [**M-4**], the categories are presented:

[M-4] Entry	Description
Mathematical	
matrix	Matrix functions
solvers	Matrix solvers and inverters
scalar	Scalar functions
statistical	Statistical functions
mathematical	Other important functions
Utility & manipulation	
standard	Standard matrices
utility	Matrix utility functions
manipulation	Matrix manipulation functions
Stata interface	
stata	Stata interface functions
String, I/O, & programming	
string	String manipulation functions
io	I/O functions
programming	Programming functions

You can click on any of the words on the left, from **matrix** to **programming**, to learn what is provided in the category. The functions will be listed in logical order with brief descriptions. Click on a function, and you will be taken to its documentation. For instance, if you clicked on **programming** and then on `timer()`, you can learn all about the function that we used in the previous chapter to measure the performance of `SpMat`. Or click on `AssociativeArray()` to learn about the functions we used to implement `SpMat`.

If you just wanted to learn more about `timer()` or `AssociativeArray()`, the quickest way would be to type `help mata timer()` or `help mata AssociativeArray()`. The point of clicking on [M-4] and then on **programming** is to find out what else is available. Mata provides functions that return the version set by the caller, return whether you should favor speed or memory in the code you write, perform advanced string parsing, work with external (global) variables, intercept the Break key, calculate hashes, make assertions, obtain the number of bytes consumed by an object, swap the contents of variables, and determine the byte order used by this computer. And there still remain 11 more categories to learn about!

One of them is `stata`. Mata provides over 60 functions for interacting with Stata, including functions for creating new Stata variables, dropping existing ones, and accessing and modifying their content. You will even find a function named `stata()` that executes Stata commands! Because Mata is part of Stata, I feel a special obligation to help you through these functions, so also see appendix A of this book. I will not cover all of them, but the manual will.

Clicking on [M-4] is the best way to learn, in an organized way, about what exists. If you already know the name of a function, however, typing `help mata` *functionname*`()` is the quickest way to get to its documentation. What if you do not know the name but are sure a function must exist to do what you need? In that case, click on [M-5]. The name of every function is displayed along with a brief description of what it does. The short descriptions are searchable. I use [M-5] at least once a day when I am writing code.

The *Mata Reference Manual* was written for people who already understand Mata, know how to program in it, and now need to find a part. It was written for you. Learn your way around it and use it.

And with that final advice, we are done. I hope you found this book helpful. As I wrote it, I tried to imagine you reading my explanations, and when you scowled at me, I doubled back and tried again. I hope that the real you did not find yourself scowling nearly as often as I imagined.

A Writing Mata code to add new commands to Stata

A.1 Overview

Mata exists for adding new features to Stata. Now that you are an expert on Mata, we can discuss how to do that. We reviewed the mechanical aspects of adding features in chapter 9 when we created new Stata command `nchooseki` by writing `nchooseki.ado`. To review, you add features to Stata by writing ado-files, and the existence of Mata does not change that. You can call Mata functions from ado-code using ado's `mata:` command, and you put the code for those functions at the bottom of the ado-file or in Mata libraries. Do the former, and the functions will be private to the ado-file. Do the latter, and the functions will be available to other ado- and do-files. You can put some functions in the ado-file and others in libraries. If we wanted to add command `xyz` to Stata, file `xyz.ado` might look like this:

── xyz.ado draft 1 ──────────

```
*! version 1.0.0
program xyz
        version 15
        ...
        mata: xyzsubr(...)
     ...
end

version 15
mata:
void xyzsubr(...)
{
        ...
        ... generalsubr(...) ...
        ...
}
end
```

── xyz.ado draft 1 ──────────

Subroutine xyzsubr() is private; its code appears in xyz.ado. Meanwhile, the code for
generalsubr() is not in the file. The jargon for this is that generalsubr() is external.
External code is stored in Mata libraries such as lmatabook.mlib.

So much for review. The subjects we are going to discuss in this appendix are

1. How should we structure the code for xyz? We can imagine one extreme where
   program xyz contains just one line—mata: xyzsubr("`0'")—and the entire logic
   for the problem appears in function xyzsubr(). The Mata function proceeds from
   receiving a single argument—a string scalar containing what the user typed—and
   produces all the results and output.

   At the other end of the spectrum, we can imagine an ado program xyz that
   itself handles the problem from start to finish and calls Mata subroutines only
   occasionally.

   And we can imagine cases in between.

2. How do we write Mata subroutines that access Stata's data, macros, scalars, and
   so on? If the subroutines are going to do anything useful, they will need to do
   that.

3. How do we write Mata subroutines that modify Stata's data, macros, and so on?
   For that matter, how do we write Mata subroutines that merely return a result
   to Stata?

4. How do we write Mata subroutines that deal elegantly with errors caused by
   users' mistaken requests? In this book, we have dealt high-handedly with errors.
   We have mostly ignored them, and we have done that even when our code might
   produce incorrect results, such as those caused by matrices that are not symmetric
   when they should be. We have let Mata automatically abort with error in other

cases, such as those caused by matrices that ought to be square but are not. In a few cases, we forced Mata to abort with error by explicitly calling Mata's _error() function. We justified our actions by arguing that the routines we wrote would be called by routines written by programmers like ourselves, and those programmers should deal with the issues before calling our routines.

We are now those programmers, and it is our responsibility to look for the causes of such problems and issue error messages designed to help the user. Mata's default traceback logs will not do because they make it appear as if the error is in our code instead of in what the user typed.

A.2   Ways to structure code

Stata does some things better than Mata and vice versa. Stata can merge datasets. You would have to write code to do that in Mata. Stata can run linear regressions. We wrote Mata code to do that in chapters 11 and 13. Stata does a lot of things well, but it falls short when solutions involve looping, complicated programming logic, and matrix operations. Such code is better written in Mata. Mata code will be easier to write and quicker to debug, and the results run faster.

The way I structure code depends on the proportion of it I envision writing in Stata and in Mata. If xyz involves solving a problem more easily coded in Stata, then I will write xyz to be the main routine—the routine in control that threads its way through the subroutines to produce the final result—and I will write each subroutine in Stata or Mata according to whichever is more natural.

If the problem is instead a natural for Mata, then I turn the design around. Stata program xyz will still be the main routine—there is no choice about that—but I will write xyz so that it quickly turns the problem over to its Mata subroutine, and it is the subroutine that will thread its way to the solution. It too will call subroutines and, just as with ado-code, those subroutines will be written in Stata or Mata.

I say that I will write xyz to quickly call its Mata subroutine, but even in programs that are more naturally written in Mata, I let the ado-code handle the parsing. Stata is outstanding at parsing Stata syntax and, once parsed, at identifying the relevant sample of observations. Stata commands syntax and marksample make doing that easy. Here is an xyz.ado for a problem that I intend to write almost entirely in Mata:

──────────────────────────────────── xyz.ado draft 2 ──────────

```
*! version 1.0.0
program xyz
        version 15

        syntax varlist(min=2 numeric) [if] [in]              ///
                        [, ALTERnate(varlist(numeric))]
        marksample touse
        markout `touse' `alternate'

        mata: xyzsubr("`varlist'", "`touse'", "`alternate'")
end

version 15

local RS real scalar
local RM real matrix
local SS string scalar
local SR string rowvector

mata:
void xyzsubr(`SS' varlist, `SS' altvarlist, `SS' touse)
{
        ...
}
end
```

──────────────────────────────────── xyz.ado draft 2 ──────────

xyz has the syntax

$$\text{xyz } \textit{varlist } [\text{if } \textit{expr}] \ [\text{in } \textit{range}] \ [, \ \underline{\text{alter}}\text{nate}(\textit{varlist}_2)]$$

varlist is required and must contain two or more variables. alternate(*varlist*$_2$) is optional. The variables in both varlists are required to be numeric. Meanwhile, xyz allows the usual if *expr* and in *range* modifiers to restrict the observations used. In this example, xyz calculates and reports a statistic based on *varlist* and, optionally, *varlist*$_2$.

The first substantive line of the ado-program parses the above syntax:

```
syntax varlist(min=2 numeric) [if] [in]              ///
                [, ALTERnate(varlist(numeric))]
```

The above two lines comprise one logical line. The forward slashes in quick succession is how you specify that physical lines are to be joined. syntax is a remarkable command. You specify the syntactical components that you wish to require or allow in nearly the form of the original syntax diagram, and syntax then stores what the user typed broken into its syntactical pieces. Those pieces are stored in Stata macros. The macros in this case will be named varlist, if, in, and alternate. syntax does that, or if what the user typed does not match what is allowed or does not provide all that is required, syntax itself issues the appropriate error messages and stops the ado-file.

The next two lines of xyz identify the sample of observations to be used, and although it is not explicit in their syntax, they do that in part by accessing the information left behind by syntax. The two lines are

```
marksample touse
markout `touse´ `alternate´
```

marksample touse creates a temporary variable in the Stata dataset that marks the observations to be used. It starts by storing the name of the temporary variable in touse. In Stata speak, 'touse' *is* the new variable. The new variable contains 0s and 1s, with the 1s marking the observations that xyz will need to use. marksample set values to 1 in observations that were not excluded by any if *expr* or in *range*, and then only if *varlist* contains no missing values. Because xyz also needs to exclude observations with missing values in $varlist_2$, I coded a second command—markout 'touse' 'alternate'. That command reset values in 'touse' to 0 in observations for which $varlist_2$ contains missing values, if it was specified.

Thus, in just three programming commands, I accomplished what Stata makes so easy: I parsed the input. I issued error messages if the user made a mistake. If not, I obtained lists of the relevant variables and identified the sample over which the statistic should be calculated. The result is that what xyz needs to do is now completely defined by 'varlist', 'alternate', and 'touse'. What xyz needs to do is going to be performed by Mata subroutine xyzsubr(), so I next called xyzsubr() and passed it the contents of the three macros:

```
mata: xyzsubr("`varlist´", "`touse´", "`alternate´")
```

Let me parse this important line for you. The mata: prefix says that what follows is to be executed by Mata, not Stata, and it says that if an error occurs in executing the Mata portion of the code, execution of the ado-file is to stop. I passed three arguments to xyzsubr(): "'varlist'", "'alternate'", and "'touse'". Understand that "'varlist'", "'alternate'", and "'touse'" are just strings of characters enclosed in double quotes. If the user typed

```
. xyz mpg weight if foreign
```

then the macros will be

```
    "`varlist´" = "mpg weight"
      "`touse´" = "__000009"
"`alternate´" = ""
```

The odd-looking __000009 is a Stata-generated temporary variable name. The variable __000009 is what contains the 0s and 1s that specify the sample. In any case, after macro expansion, the line coded in the ado-file, which is

```
mata: xyzsubr("`varlist´", "`touse´", "`alternate´")
```

is interpreted as if it read

```
mata: xyzsubr("mpg weight", "__000009", "")
```

That is the line that Stata will pass to Mata for execution. The Mata code that receives the call reads

```
void xyzsubr(`SS´ varlist, `SS´ altvarlist, `SS´ touse)
{
        . . .
}
```

Thus, the Mata string scalar variables **varlist**, **altvarlist**, and **touse** will contain

```
   varlist = "mpg weight"
altvarlist = ""
     touse = "__000009"
```

xyzsubr() will take those three string values and produce the desired result. We are going to fill in the code for **xyzsubr()** in the sections that follow. Here is a preview:

1. We will make matrices D and possibly A. Matrix D will contain the relevant observations on the variables in **varlist**. A will contain the relevant observations on **altvarlist** if **altvarlist!=""**. In the example, D will be $N \times 2$. It will have N rows and two columns, one each for **mpg** and **weight**. The N observations will be those for which Stata variable **__000009!=0**. Meanwhile, A will be 0×0 because Mata variable **altvarlist==""**.

2. The purpose of **xyz** is to calculate a statistic. We will calculate the statistic based on D and A. It happens that the statistic can be meaningfully calculated only when there are no repeated variables in D. This means that if the user typed **xyz mpg weight mpg**, then **xyz** will need to issue an error message and stop.

3. If the user did not repeat a variable, we will calculate and display the statistic's value and store it in **r(statistic)**.

We are not going to get lost in math. The statistic that we will calculate is

```
statistic = 2
```

That line will be the stand-in for a more complicated calculation involving D and, if it is specified, A. To help you see where we are going, here is how I envision our not-yet-written command will look in action:

```
. xyz mpg weight mpg if foreign
  repeated variables not allowed
r(498);

. xyz mpg weight if foreign

        statistic = 2

. return list
scalars:
            r(statistic) =    2


. _
```

A.3   Accessing Stata's data from Mata

Mata provides six functions for accessing the data that Stata has stored in memory. They are

Function name	Purpose
st_data()	Copy numeric data from Stata into Mata
st_sdata()	Copy string data from Stata into Mata
st_view()	Access Stata's copy of numeric data
st_sview()	Access Stata's copy of string data
_st_data()	(seldom used, numeric data)
_st_sdata()	(seldom used, string data)

We will ignore the last two functions because I have never found a use for them. You can type `help mata _st_data()` to learn more about them if you are curious.

The other four functions are enormously useful. They are how Mata programmers access Stata's data. Notice that the four functions occur in two pairs, one for accessing Stata's numeric variables, such as `mpg` and `weight`, and another for accessing string variables. Mata does not allow mixed-type matrices, so you cannot put numeric and string variables in the same matrix. You instead put them in separate matrices if you have both types, and Mata provides separate functions for doing that. `st_data()` and `st_sdata()` copy numeric and string data from Stata into Mata. Their base syntax is

```
numericresult = st_data(i, j)
```

```
stringresult = st_sdata(i, j)
```

Arguments i and j specify the observations and variables to be returned. There are lots of variations on how they can be specified. Both st_data() and st_sdata() work the same way. What is said about st_data() below applies equally to st_sdata().

1. i and j can be specified as integers.
   st_data(2,3) returns the value of the second observation of the third variable in the Stata dataset. Perhaps 17 would be returned. Nobody uses st_data() in this way, but they could.

2. j can be specified as a variable name.
   st_data(2, "mpg") returns the second observation on mpg. Returned might be the same value that was returned when we specified j as 3. Nobody uses st_data() in this way either.

3. i can be missing value.
   This is useful. st_data(., "mpg") returns an $N \times 1$ column vector of all the observations on mpg.

4. j can be a vector.
   This is even more useful. st_data(., ("mpg", "weight")) returns an $N \times 2$ matrix containing all the observations on mpg and weight.

i and j can be specified in other ways, too. You can learn all the ways by typing help mata st_data(). I am going to show you how functions are mostly used and, even in syntax diagrams, not burden you with all their syntaxes and features. I am also not going to mention this again, so do not assume that a feature is absent just because I do not tell you about it.

Concerning the code for xyzsubr() that we need to write, st_data() takes us halfway to producing the D matrix that we need. We could code

```
D = st_data(., tokens(varlist))
```

I say halfway because D produced in this way would contain all the observations in the dataset, and we want all the observations for which variable __000009!=0. I will show you how we can fix that, but first I need to explain why the line reads

```
D = st_data(., tokens(varlist))
```

and not

```
D = st_data(., varlist)
```

Variable varlist, you will recall, is a string scalar in the Mata function that we are writing. It contains a space-separated list of variable names, such as "mpg weight". st_data(i,j) requires that j be a vector when more than one variable is being specified. st_data() does not want to see "mpg weight", it wants to see ("mpg", "weight"), a 1×2 vector. Mata's tokens() function converts space-separated lists into such vectors.

I said that we are halfway to the D matrix we need, but we also need to restrict the observations to the subset for which __000009!=0. st_data() and st_sdata() provide a syntax for that. The variable name containing the nonzero values that specify the observations to be selected can be specified as a third argument:

```
numericresult = st_data(., j, tousevar)
```

```
stringresult = st_sdata(., j, tousevar)
```

With three arguments, the functions return all observations on variables j such that the Stata variable name recorded in the third argument is not equal to 0. That is exactly what we need. Variable touse in xyzsubr(), I remind you, is the string scalar containing the variable name __000009. We could code

```
D = st_data(., tokens(varlist), touse)
```

Before we use the above solution, however, there is another we should consider. The problem with st_data() and st_sdata() is that they copy data from Stata into Mata, and copies consume memory. The memory is not an issue for small datasets, but it can be important with larger ones. st_view() and st_sview() are variations on st_data() and st_sdata() that solve the memory problem. They do just what st_data() and st_sdata() do, but the matrices they construct are views onto the Stata data instead of being a copy. Their syntax is somewhat different, too. Instead of typing

```
D = st_data(., tokens(varlist), touse)
```

you type

```
st_view(D, ., tokens(varlist), touse)
```

That is, what previously appeared on the left side of assignment now appears as the lead argument of the view function. Other than that, the syntax of the view functions mimics that of the data functions:

```
st_view(numericresult, i, j)
```

```
st_sview(stringresult, i, j)
```

```
st_view(numericresult, ., j, tousevar)
```

```
st_sview(stringresult, ., j, tousevar)
```

Views are different from copies, but after construction, view matrices work as if they were ordinary matrices. The memory the views consume, however, is trivial compared with the copies produced by st_data() and st_sdata(). Views have a corresponding disadvantage. The time it takes to access the elements of views is a little greater. If matrix R is created by st_data() and matrix V, by st_view(), but they are otherwise the same matrix, it takes longer to access elements of V than it does to access elements of R. Do not make too much of this, because in most cases, you will not notice it. If, however, you were to calculate R*b or V*b millions of times for a column vector b, you will notice the difference.

Which should you use? Mata has a setting about whether it should favor memory or speed of execution. Mata's built-in function favorspeed() returns the setting. In the case of the xyzsubr() example, we could code

```
if (favorspeed()) D = st_data(., tokens(varlist), touse)
else                   st_view(D,  ., tokens(varlist), touse)
```

Most programmers do not bother to do this because most users never set whether they prefer speed or space; thus, favorspeed() is always false, which is the default Stata ships with. In cases where we at StataCorp have been uncertain about whether to use st_data() or st_view(), we have written out code in just the way shown and we have done timings both ways. Then we chose once and for all whether to substitute st_data() or st_view() for the two lines. I know of only a few instances in which we finally settled on st_data() because of the speed penalty of st_view(), and those instances involved massive amounts of calculation.

To summarize, when writing an ado-file that uses Mata, the usual and best approach is to

1. Use syntax in the ado-portion of the code to parse what the user types.

2. Use marksample and perhaps markout in the ado-portion of the code to create a 'touse' variable.

3. Pass "'touse'" to the Mata function. Declare the argument that receives the variable name as string scalar touse.

4. Construct matrices and vectors using
        st_view(*name*, ., j, touse)
   or
        name = st_data(., j, touse)
   It is generally better to use st_view().

I told you that view matrices work just like regular matrices, but there are a few exceptions where they do not work at all. There are Mata functions and operators that, given a view matrix, issue the error "view found where array required", r(3103). Mata issues the error if you attempt to obtain the Kronecker product and one of the matrices is a view. Mata also issues the error if you attempt to use function _transpose() on a view. The writeups in the *Mata Reference Manual* mention when views are not allowed.

Just because Mata says "view found where array required" does not mean you cannot use views. You can code around the problem, and that is sometimes a better solution. In these cases, I suggest you still consider whether a view might not be the better solution because you can usually work around the problem. In spmat3.mata, discussed in chapter 18, we had code that used _transpose(). It read

```
`Regular´ R_SxR(`Sparse´  A, `boolean´ tA,
              `Regular´ B, `boolean´ tB)
{
      `Regular´ R

      if (tB) _transpose(B)
      R = A.R_SxR(tA, B)
      if (tB) _transpose(B)
      return(R)
}
```

This code will not work when B is a view and tB is not equal to 0. That is a considerable shortcoming because the user of the sparse matrix system might have good reason to want to specify a view. If we complicated its logic, the code could be made to work with views without sacrificing the advantages of using _transpose() when B is not a view. The following draft uses Mata's isview() function to determine whether B is a view:

```
`Regular´ R_SxR(`Sparse´  A, `boolean´ tA,
              `Regular´ B, `boolean´ tB)
{
      `Regular´ R

      if (tB) {
              if (isview(B)) R = A.R_SxR(tA, B´)
              else {
                      if (tB) _transpose(B)
                      R = A.R_SxR(tA, B)
                      if (tB) _transpose(B)

              }
      }
      else    R = A.R_SxR(tA, B)
      return(R)
}
```

In any case, there is no issue caused by using views in the code for xyz.ado, and so we are going to use it. The updated draft reads

```
                                                    ─── xyz.ado draft 3 ───────
*! version 1.0.0
program xyz
        version 15

        syntax varlist(min=2 numeric) [if] [in]          ///
                    [, ALTERnate(varlist(numeric))]
        marksample touse
        markout `touse´ `alternate´

        mata: xyzsubr("`varlist´", "`touse´", "`alternate´")
end

version 15

local RS real scalar
local RM real matrix
local SS string scalar
local SR string rowvector

mata:
void xyzsubr(`SS´ varlist, `SS´ altvarlist, `SS´ touse)
{
        `RM´  D, A

        pragma unset D
        st_view(D, ., tokens(varlist), touse)

        if (altvarlist!="") {
                pragma unset A
                st_view(A, ., tokens(altvarlist), touse)
        }
        ...                                    // <-- we are here
}
end
                                                    ─── xyz.ado draft 3 ───────
```

Sharp-eyed readers will notice the two pragmas in the code. They are there because Mata would otherwise complain, if we set `matastrict` on, that D and A may be set before they are used. `matastrict` does not understand that the two `st_view()` statements set D and A because they do not appear on the left of an assignment statement.

A.4   Handling errors

Mata does not handle errors gracefully. As far as Mata is concerned, errors should not happen in well-written code, and if they do, Mata aborts execution and presents a traceback log that will help the competent programmer find his or her error.

Stata has an entirely different attitude. Users make mistakes, and when they do, not only does execution need to stop (Mata and Stata agree on that), but a tolerant, helpful error message should be presented that in effect invites the user to try again.

Your job as a programmer of Mata subroutines for ado-files is to write code to present these tolerant and helpful error messages and to cancel further execution without the presentation of a traceback log. The key to doing the last part is Mata's `exit(`*rc*`)` function. Before you do that, you use Mata's `errprintf()` function to display the error message. `errprintf()` is like `printf()`, except that `errprintf()` marks the output as error output so that it is displayed in Stata's error style, which usually means that it appears in red.

We are going to do just that in `xyzsubr()`. You may remember that there can be no repeated variable names in `varlist` if the statistic we will calculate is to be meaningful. We need to determine if repeated variables are specified, and if so, issue the error message and exit. To remind you, the draft of function `xyzsubr()` at this point is

```
void xyzsubr(`SS´ varlist, `SS´ altvarlist, `SS´ touse)
{
        `RM´ D, A

        pragma unset D
        st_view(D, ., tokens(varlist), touse)

        if (altvarlist!="") {
                pragma unset A
                st_view(A, ., tokens(altvarlist), touse)
        }
        ...                                     //  <-- we are here
}
```

We will present the error message and exit only if there are repeated variable names. I will show you how we will do that and then explain:

```
cv = tokens(varlist)´
if (length(uniqrows(cv)) != length(cv)) {
        error message and exit go here
}
```

In the code, `cv` is the same as `varlist`, except the names are stored in a column vector instead of a scalar containing names separated by spaces. The `if` statement asks whether the number of unique names in `cv` equals the number of names in `cv`. Mata built-in function `uniqrows(`*x*`)` returns, well, the unique rows of *x*. It returns *x* with duplicate rows removed. All that is left to do is add the error message and `exit()` with the appropriate return code:

```
cv = tokens(varlist)´
if (length(uniqrows(cv)) != length(cv)) {
        errprintf("  repeated variables not allowed")
        exit(498)
        // NotReached
}
```

The updated version of `xyz.ado` is

```
————————————————————————————————————————— xyz.ado draft 4 ———————
*! version 1.0.0
program xyz
        version 15

        syntax varlist(min=2 numeric) [if] [in]              ///
                        [, ALTERnate(varlist(numeric))]
        marksample touse
        markout `touse' `alternate'

        mata: xyzsubr("`varlist'", "`touse'", "`alternate'")
end

version 15

local RS real scalar
local RM real matrix
local SS string scalar
local SR string rowvector
local SC string colvector

mata:
void xyzsubr(`SS' varlist, `SS' altvarlist, `SS' touse)
{
        `RM'  D, A
        `SC'  cv

        pragma unset D
        st_view(D, ., tokens(varlist), touse)

        if (altvarlist!="") {
                pragma unset A
                st_view(A, ., tokens(altvarlist), touse)
        }

        cv = tokens(varlist)'
        if (length(uniqrows(cv)) != length(cv)) {
                errprintf("  repeated variables not allowed")
                exit(498)
                // NotReached
        }
        //                                        <-- we are here
}
end
————————————————————————————————————————— xyz.ado draft 4 ———————
```

Now when users type duplicated variable names, they will see

```
. xyz mpg weight mpg if foreign
  repeated variables not allowed
r(498);
```

A.5   Making the calculation and displaying results

Recall how **xyz** is to work when the user does not make an error:

```
. xyz mpg weight if foreign

        statistic = 2

. return list
scalars:
          r(statistic) =    2


. _
```

We have structured the code so that ado-program **xyz** parses the input, identifies the sample, and calls Mata function **xyzsubr()**. Mata function **xyzsubr()** is to calculate the statistic, display its value, and return the value **xyz**. We have a partial draft of **xyzsubr()** that handles the setup for doing this. It reads

```
void xyzsubr(`SS´ varlist, `SS´ altvarlist, `SS´ touse)
{
        `RM´  D, A
        `SC´  cv

        pragma unset D
        st_view(D, ., tokens(varlist), touse)

        if (altvarlist!="") {
                pragma unset A
                st_view(A, ., tokens(altvarlist), touse)
        }

        cv = tokens(varlist)´
        if (length(uniqrows(cv)) != length(cv)) {
                errprintf("  repeated variables not allowed")
                exit(498)
                // NotReached
        }
        //                                      <-- we are here
}
```

Now we need to add the code to make the calculation and display the result. We have already agreed to use the formula

```
statistic = 2
```

Of course, the true statistic is not really 2. The formula is a stand-in for the code that will produce the true value based on the matrices **D** and **A**. This stand-in is no different from how we have written code throughout this book. We start with second-rate formulas and code better formulas later. I admit that this formula does not even make an attempt to calculate the statistic correctly, but so what? You could call it a third-rate formula if it would make you feel better.

Now we need to display the result stored in `statistic`. `printf()` is how you display output in Mata. After the calculation `statistic = 2`, we will add the lines

```
printf("\n")
printf("{txt:    statistic = }{res:%9.2f}\n", statistic)
```

You can learn more about `printf()` by typing `help mata printf()`.

The updated `xyzsubr()` now reads

```
void xyzsubr(`SS´ varlist, `SS´ altvarlist, `SS´ touse)
{
        `RM´  D, A
        `SC´  cv

        pragma unset D
        st_view(D, ., tokens(varlist), touse)

        if (altvarlist!="") {
                pragma unset A
                st_view(A, ., tokens(altvarlist), touse)
        }

        cv = tokens(varlist)´
        if (length(uniqrows(cv)) != length(cv)) {
                errprintf("  repeated variables not allowed")
                exit(498)
                // NotReached
        }

        statistic = 2

        printf("\n")
        printf("{txt:    statistic = }{res:%9.2f}\n", statistic)

        //                                      <-- we are here
}
```

A.6   Returning results

The final step in writing `xyz.ado` is to add the code to `xyzsubr()` to return the value of `statistic` to its caller, `xyz`, so that `xyz` can return `r(statistic)` to its users. There are seven commonly used ways to return results from Mata to Stata.

- To return a Mata string scalar.
  - (1) store it as a named Stata local macro that the ado-file can access
  - (2) store it as a named Stata string scalar that the ado-file can access
  - (3) store it as a Stata `r()` macro that the ado-file can access

- To return a Mata real scalar:
  - (4) store it as a named Stata numeric scalar that the ado-file can access
  - (5) store it as a Stata `r()` scalar that the ado-file can access

- To return a Mata real vector or matrix:
  - (6) store it as a named Stata matrix that the ado-file can access
  - (7) store it as a Stata `r()` matrix that the ado-file can access

I recommend returning results in `r()` in all cases. When ado-files call other ado-files, results are returned in `r()`. Why should Mata subroutines behave differently? Here is how to store in `r()` from Mata:

- To return Mata string scalar s, call `st_global("r(`name`)", `s`)`

- To return Mata real scalar r, call `st_numscalar("r(`name`)", `r`)`

- To return Mata vector or matrix M, call `st_matrix("r(`name`)", `M`)`

Because we need to return a Mata real scalar, we will use `st_numscalar()`:

```
st_numscalar("r(result)", statistic)
```

`xyzsubr()` will now read

```
void xyzsubr(`SS' varlist, `SS' altvarlist, `SS' touse)
{
        `RM'  D, A
        `SC'  cv

        pragma unset D
        st_view(D, ., tokens(varlist), touse)

        if (altvarlist!="") {
                pragma unset A
                st_view(A, ., tokens(altvarlist), touse)
        }

        cv = tokens(varlist)'
        if (length(uniqrows(cv)) != length(cv)) {
                errprintf("  repeated variables not allowed")
                exit(498)
                // NotReached
        }

        statistic = 2

        printf("\n")
        printf("{txt:    statistic = }{res:%9.2f}\n", statistic)

        st_numscalar("r(result)", statistic)
}
```

Note that I stored the statistic in `r(result)` even though the ultimate goal is to return `r(statistic)`. I could have stored it in `r(statistic)`, but I stored it in `r(result)` to emphasize to you that `r(result)` is the value that `xyzsubr()` is returning to its caller, ado-program `xyz`, and not the result that `xyz` will return to its caller, the user of the command. This is the same situation as with all ado-files. Think about an ado-file `simple.ado` that does not even use Mata. Assume that `simple` is r-class, and someplace along the way, it uses Stata's `summarize` command to calculate a mean. It wants to return that value to its users as `r(y_mean)`. The code would read

```
program simple, rclass
        version 15

        syntax ...
        marksample touse
        .
        .
        quietly summarize `lhsvar' if `touse'
        return scalar y_mean = r(mean)
        .
        .
end
```

summarize leaves behind r(mean) and other r() results for its caller's use. In the code, simple adds the value stored in r(mean) to its returned results as r(y_mean) using the return command. Had simple wanted to call its returned result r(mean), it would still have had to add it using the return command. The command would have been

```
return scalar mean = r(mean)
```

Program xyz works the same way. Program xyz called a subroutine that happened to be written in Mata, but that makes no difference. The Mata subroutine leaves behind r() results that its caller can use (or ignore) as it pleases. In the code below, xyz stores r(result) in its return value r(statistic). Had xyzsubr() returned r(statistic), xyz would still need to add r(statistic) to its returned results.

Ado-program xyz is now

```
program xyz, rclass
        version 15

        syntax varlist(min=2 numeric) [if] [in]              ///
                       [, ALTERnate(varlist(numeric))]
        marksample touse
        markout `touse´ `alternate´

        mata: xyzsubr("`varlist´", "`touse´", "`alternate´")
        return scalar statistic = r(result)
    end
```

We are done with the xyz file, but I was sloppy when I added the line "return scalar result = statistic" in xyzsubr(). I should have added the line "st_rclear()" above it. I often forget and it seldom matters. st_rclear() clears r(). Mata does not clear r() automatically like Stata does. Because it is not automatic, had the caller of xyzsubr() used summarize just before calling xyzsubr(), then later, after xyzsubr() returned, r() would have contained summarize's returned results as well as xyzsubr()'s r(result). The extra results do not matter as long as the caller does not use return add to add all values currently stored in r() to the r() that it will return. If it did, there would be more stored in r() (and thus returned) than intended.

Here is the final version of xyz.ado, including the st_rclear() that I previously omitted:

── xyz.ado final ───────────

```
*! version 1.0.0
program xyz
        version 15

        syntax varlist(min=2 numeric) [if] [in]              ///
                        [, ALTERnate(varlist(numeric))]
        marksample touse
        markout `touse' `alternate'

        mata: xyzsubr("`varlist'", "`touse'", "`alternate'")
        return scalar statistic = r(result)
end

version 15

local RS real scalar
local RM real matrix
local SS string scalar
local SR string rowvector
local SC string colvector

mata:
void xyzsubr(`SS' varlist, `SS' altvarlist, `SS' touse)
{
        `RM'   D, A
        `SC'   cv
        `RS'   statistic

        pragma unset D
        st_view(D, ., tokens(varlist), touse)

        if (altvarlist!="") {
                pragma unset A
                st_view(A, ., tokens(altvarlist), touse)
        }

        cv = tokens(varlist)'
        if (length(uniqrows(cv)) != length(cv)) {
                errprintf("  repeated variables not allowed")
                exit(498)
                // NotReached
        }

        statistic = 2

        printf("\n")
        printf("{txt:   statistic = }{res:%9.2f}\n", statistic)

        st_rclear()
        st_numscalar("r(result)", statistic)
}
end
```

────────────────────────────────────── xyz.ado final ───────────

I have labeled the above code as final, but it is not final. We still have to substitute the correct calculation for the line "`statistic = 2`". I ought to put a double-bang (!!) next to the line so that I do not forget, because we are not going to do that here.

A.7 The Stata interface functions

We have discussed only a handful of the Stata interface functions that Mata provides. I will tell you about the rest of them, but not in great detail. If it is detail you seek, see the *Mata Reference Manual*. As you read about the functions, type `help mata` *functionname*() for any that interest you.

A.7.1 Accessing Stata's data

We covered the functions for accessing Stata's data. They are

Function name	Purpose
st_data()	copy numeric data from Stata into Mata
st_sdata()	copy string data from Stata into Mata
st_view()	access Stata's copy of numeric data
st_sview()	access Stata's copy of string data
_st_data()	(seldom used, numeric data)
_st_sdata()	(seldom used, string data)

We covered the functions, but I have a warning for those of you who use `st_view()` or `st_sview()`. It is your responsibility to reestablish views if you drop variables, rearrange their order, or drop observations. Consider the following example:

```
st_view(V, ., ("a", "c"))
.
.
.
st_dropvar("b")
```

After `st_dropvar()`, the view V may need to be reestablished. It needs to be reestablished if the order of the variables in the dataset is a, b, and c because then, after dropping variable b, the second column of V will become whatever variable comes after c, presumably d, but perhaps nothing. Views base the variables they view on variable numbers. When you established the view by coding

```
st_view(V, ., ("a", "c"))
```

Mata recorded that V is a view onto variables 1 and 3 of the Stata dataset. After you dropped b, variable c became variable 2, but the view does not know that. V is still a view onto variables 1 and 3. For V to be a view onto a and c, you need to reestablish it, which you do by repeating the command that originally created it:

```
st_view(V, ., ("a", "c"))
```

The same issue arises with dropped observations.

A.7.2   Modifying Stata's data

The functions for modifying the values recorded in Stata's data mirror those for accessing the values. They are

Function name	Purpose
st_store()	like st_data(), but stores
st_sstore()	like st_sdata(), but stores
st_view()	access Stata's copy of numeric data
st_sview()	access Stata's copy of string data
_st_store()	like _st_data(), but stores
_st_sstore()	like _st_sdata(), but stores

These modify functions mimic the syntaxes of their corresponding access functions.

Notice that st_view() and st_sview() are both access and modify functions. They modify values when the matrix they create appears on the left of assignments. Be sure to specify subscripts explicitly in such cases. The following code changes each value of mpg in Stata's data to be its reciprocal, gallons per mile:

```
st_view(v, ., "mpg")
v[.] = 1 :/ v
```

If you omitted the subscripts and typed v = 1:/v, v changes from being a view to being a regular matrix. The Stata data would be unchanged, although the new, regular matrix v would contain gallons per mile. This is explained in section 5.11.1.

A.7.3   Accessing and modifying Stata's metadata

Metadata refers to data associated with Stata's data, such as %fmts and variable labels. Even the variable names are metadata, as are the number of observations and number of variables.

Stata's metadata are

- For the data as a whole:
    - # of variables
    - # of observations
    - dataset characteristics
      date and time last written
    - filename if any
    - whether changed since last saved

- For each variable of the data:
    - variable name
    - storage type (`byte`, `int`, ...)
    - display format (`%fmt`)
    - value-label name
    - variable label
    - variable characteristics

- For each value-label name:
    - numeric values
    - mapped value

The following functions make all the above accessible and allow modification of formats, labels, and characteristics:

Function name	Purpose
st_nvar()	# of variables
st_nobs()	# of observations
st_global()	dataset characteristics
c()	date and time last written
c()	filename
st_updata()	whether changed since last saved
st_varindex()	variable numbers of variable names
st_varname()	variable names of variable numbers
st_vartype()	storage type variable
st_isnumvar()	whether variable is numeric
st_isstrvar()	whether variable is string
st_varformat()	display format of variable
st_varlabel()	variable label of variable
st_varvaluelabel()	value-label name of variable
st_global()	variable characteristics
st_vlexists()	value-label definition (suite of functions)

Function c() appears twice in the above table, and the st_global() function that appears will appear in other tables. There is no mistake. Mata's functions are sometimes overloaded, and the arguments you specify determine what is returned or changed.

You may someday need the name of every Stata variable in Stata's data, and you will search without success for the function that does that. st_varname() will return the full list if you code

```
varnames = st_varname(1..st_nvar())
```

A.7.4   Changing Stata's dataset

Stata's dataset is Stata's data and metadata, taken together. Some things cross the line between being data and metadata, such as the number of observations, the number of variables, and the variable names. Renaming variables and adding or dropping variables or observations are considered changes to Stata's dataset.

The functions for changing Stata's dataset are

Function name	Purpose
st_varrename()	rename variable
st_addvar()	add variable to data
st_dropvar()	drop variable from data
st_keepvar()	keep variable in data
st_addobs()	add observations to data
st_dropobsif()	drop selected observations from data
st_keepobsif()	keep selected observations in data
st_keepobsin()	keep specified observations in data

A.7.5   Accessing and modifying Stata macros, scalars, matrices

Stata provides macros, scalars, and its own concept of matrices separate from that of Mata, and these things are separate from Stata's dataset. They provide the variables for Stata's ado-programming.

You can access and change the contents of Stata's macros, scalars, and matrices using the following functions:

Function name	Purpose
st_global()	Stata's global macros
st_local()	Stata's local macros
st_numscalar()	Stata's numeric scalars
st_strscalar()	Stata's string scalars
st_matrix()	Stata's matrices
st_rclear()	clear Stata's r()
st_global()	Stata's r() macros
st_numscalar()	Stata's r() numeric scalars
st_strscalar()	Stata's r() string scalars
st_matrix()	Stata's r() matrices
st_eclear()	clear Stata's e()
st_global()	Stata's e() macros
st_numscalar()	Stata's e() numeric scalars
st_strscalar()	Stata's e() string scalars
st_matrix()	Stata's e() matrices
st_scalar()	clear Stata's s()
st_global()	Stata's s() macros

A.7.6   Executing Stata commands from Mata

We have discussed how Stata can call Mata functions using Stata's `mata:` command. The reverse capability is also provided. Mata can execute Stata commands using either the `stata()` or `_stata()` function:

Function name	Purpose
stata()	execute specified Stata command, abort with error if $rc \mathrel{!}= 0$, and otherwise return void
_stata()	execute specified Stata command and return rc

These functions differ only in how they treat errors that occur while the Stata command is being executed. `stata()` aborts with error, meaning it stops execution and presents a traceback log. `_stata()` returns a real scalar equal to the return code. Return codes equal 0 when no errors occurred. Understand that even when errors do occur, the Mata code still regains control, and it is up to that code to determine how to deal with the problem.

_stata() is the command you want to use in serious code. stata() is useful in do-files and the like for private use. Both commands allow one to three arguments. We will discuss _stata(). Its syntax is

```
rc = _stata(cmd)


rc = _stata(cmd, suppressoutput)


rc = _stata(cmd, suppressoutput, suppressmacroexpansion)
```

cmd is a string scalar containing the line that Stata is to execute. The other arguments are real scalars, with 0 meaning false and nonzero meaning true. The return value rc is a real scalar.

If you coded

```
rc = _stata("regress mpg weight displ")
```

then Stata would execute the regression command and show the output. Presumably, _stata() would return 0. If you coded

```
rc = _stata("regress mpg weight displ", 1)
```

then Stata would execute the regression command but show no output. Not even error messages will be displayed.

The third argument works the same way but affects whether Stata's usual macro expansion is to be performed. In rare instances, suppressing macro expansion can be useful, but the third argument is mostly specified for speed execution time. Even so, few programmers specify it because macro expansion is so fast.

Here is a Mata function that runs a linear regression using Stata and returns the coefficient vector:

```
real colvector stregress(string scalar lhs, string rowvector rhs)
}
        real scalar    rc
        string scalar cmd

        cmd = "regress " + lhs + " " + invtokens(rhs)
        if ((rc = _stata(cmd, 1))) {
                errprintf("  could not run regression\n")
                errprintf("  Here's the output:")
                (void) _stata(cmd)
                exit(rc)
        }
        return(st_matrix("e(_b)"))
}
```

I have a more useful example. You may have noticed that in the presentation of Stata's metadata above, I omitted mentioning Stata's dataset label. I omitted it because there is not an st_*() function to return it because of an oversight by StataCorp. Stata can get the dataset label, however, so we could redress the omission. We could write a new Mata function using _stata() to execute the appropriate Stata commands. That function could be

```
string scalar st_datalabel()
{
        string scalar   macname, label

        macname = st_tempname()
        label   = ""

        if (_stata("local " + macname + " : data label")==0)
                label = st_local(macname)
                st_local(tempname, "")   // delete macro tempname
        }
        return(label)
}
```

I should mention that st_tempname() is a Stata interface function that I have not told you about. It returns a temporary name, the same as by Stata's tempname command.

A.7.7   Other Stata interface functions

There are lots of Stata interface functions besides st_tempname() that I have not told you about. To find out about all of them, type help mata, click on [M-4], and click on stata. Do that, and you will be looking at all the functions.

B Mata's storage type for complex numbers

B.1 Complex values

In chapter 3, we discussed Mata's storage types but only touched on complex numbers. Complex values were slighted because most Mata programmers do not use them. The following example should reassure you that Mata can perform calculations on the complex plane. 1i is how you write i in Mata.

```
: // -------------------------------- multiplication and division
: (1+1i)*(2-1i)
  3 + 1i

: (3+1i)/(2-1i)
  1 + 1i

:
: // powers
: (1+1i)^2
  2i

: ((2i)^(1/2), sqrt(2i))
              1        2
  ┌──────────────────────┐
1 │  1 + 1i    1 + 1i     │
  └──────────────────────┘

: z = (1+1i)^(2-1i)
```

```
: z, z^(1/(2-1i))
```

	1	2
1	1.49001412 + 4.12574447i	1 + 1i

```
:
: // ---------------------- natural logarithm and exponentiation
: ln(1+1i)
  .34657359 + .785398163i
: exp(ln(1+1i))
  1 + 1i

:
: // ------------------------ length, real, imaginary, conjugate
: abs(1+1i)
  1.414213562
: Re(1+2i)
  1
: Im(1+2i)
  2
: conj(1+2i)
  1 - 2i

:
: // ---------------------------------------------- inversion
: z = 1+1i
: luinv(z)
  .5 - .5i
: 1/z
  .5 - .5i
: Z = (2+1i, 2-1i \ 2+1i, 2+1i)
: luinv(Z)
```

	1	2
1	-.5i	.4 + .3i
2	.5i	-.5i

```
: Z*luinv(Z)
[Hermitian]
```

	1	2
1	1	
2	0	1

B.2   Complex values and literals

In Mata, you write complex numbers (literals, in computer speak) as

$$realpart + imaginarypart\,\text{i}$$

For instance,

```
5 + 6i
5 - 6i
-5 + 6i
-5 - 6i

1 + 1i

1.234 - 2.456i

1.02e-2 + 1e-2i
```

If a number is purely imaginary, you can omit the real part:

```
6i
1i
-1i
```

You write 1i to mean the square root of -1

If you want specify a real number and have it be treated as complex, you must either specify its zero imaginary part by adding 0i to it or use Mata's cast-to-complex C() function:

```
-1+0i
C(-1)
```

When calculating complex results, it is important that any mathematically real values be specified or stored as complex. Mata treats the literal -1 as real, and thus the square root of -1 is missing. The square root of -1+0i, however, is 1i.

```
: sqrt(-1)          // real argument
  .
: sqrt(-1+0i)       // complex argument
  1i
: sqrt(C(-1))       // complex argument
  1i
:
: x = -1            // real x
: z = -1 +0i        // complex z; z = C(-1) would also be complex
: sqrt(x)
  .
: sqrt(z)
  1i
```

Complex values are stored as a pair of real values. You can express the real and imaginary parts in any of the ways you can express real values, such as

```
1 + 1i
1.234 - 2.456i
1.02e-2 + 1e-2i
1.0x-4ai
```

The real and imaginary parts of complex values can range over the allowed values of reals, namely, $-8.988e{-}307$ to $8.988e{+}307$. The number closest to 0 without being 0 is roughly $1e{-}323$.

Complex variables can contain missing values such as ., .a, .b, ..., .z. You cannot set the real and imaginary parts to missing separately. Either the entire complex number equals a single missing value or it contains nonmissing real and imaginary parts. A complex value cannot be 1+.i and .+2i, although 1+.i and .+2i are valid expressions. They are valid expressions that are evaluated and, just as 1+. and .+2 evaluate to missing value (.), so do 1+.i and .+2i.

```
: (1+1i) + .        // expression evaluating to missing
  .

: . + 2i            // expression evaluating to missing
  .

: 2 + .i            // expression evaluating to missing
  .
```

B.3   Complex scalars, vectors, and matrices

You can create scalars, vectors, and matrices containing complex values, such as

```
: z = (86+12i, 13-2i)
: Z = (z \ 12+2i, 22+14i)
: z
                  ·1              2
    1 |    86 + 12i      13 - 2i |

: Z
                    1              2
    1 |    86 + 12i      13 - 2i |
    2 |    12 - 3i       22 + 14i |
```

Mata's transpose operator becomes the conjugate transpose when scalars, vectors, and matrices are complex:

```
: Z´
```

	1	2
1	86 - 12i	12 + 3i
2	13 + 2i	22 - 14i

Matrix Z' is Z transposed and the elements replaced with their complex conjugate. The complex conjugate of $a + bi$ is $a - bi$.

```
: z = 1 + 2i
: conj(z)
  1 - 2i
```

A real matrix is symmetric if $R = R'$. A complex matrix is correspondingly Hermitian if $Z = Z'$, where transpose means conjugate transpose. Hermitian matrices are the complex analog of real symmetric matrices. If R is real, then $R'R$ is symmetric. Correspondingly, if Z is complex, then $Z'Z$ is Hermitian:

```
: H = Z´Z
: H
[Hermitian]
```

	1	2
1	7693	
2	1316 + 94i	853

```
: H[1,2]
  1316 - 94i
: H[2,1]      // note that Im(H[2,1]) == -Im(H[1,2])
  1316 + 94i
```

Mata noted that H is Hermitian when it displayed the matrix, and it did not print the $(1, 2)$ element because it is equal to the conjugate of the $(2, 1)$ element.

Pure symmetry plays no role with complex matrices. If you make a purely symmetric complex matrix, however, Mata will label it `symmetriconly`:

```
: S
[symmetriconly]
                 1              2
      ┌─────────────────────────────┐
   1  │  1 + 2i                      │
   2  │  3 - 1i              7        │
      └─────────────────────────────┘

: S[1,2]
   3 - 1i
: S[2,1]
   3 - 1i
```

We use the terms *pure symmetry* and *symmetriconly* because complex analysts sometimes refer to Hermitian matrices as complex symmetric matrices and then will even drop the word *complex*. Mata's `issymmetric()` function follows this habit. If matrix Z is complex, `issymmetric(Z)` returns 1 if the matrix is Hermitian and 0 otherwise. It does this so that programmers can write code that will work with both real and complex matrices. They can ask `issymmetric(A)` whether A is real or complex and expect the mathematically appropriate response.

B.4 Real, complex, and numeric eltypes

You need to distinguish between the mathematical meanings and computer storage-type meanings of the words *real* and *complex*. We will distinguish between them in this chapter by writing `real` and `complex` for their computer storage-type meanings and reserve "real" and "complex" in ordinary typeface for their mathematical meanings.

A `real` matrix contains real values.

A `complex` matrix contains real values, complex values, or a mix.

Numeric variables in Mata are either `real` or `complex`. Mata allows variables to be declared `numeric`, meaning that the variable is not restricted to be one or the other.

Let's imagine three related functions:

```
real    matrix  foo1(real    matrix A)
complex matrix  foo2(complex matrix A)
numeric matrix  foo3(numeric matrix A)
```

Function `foo1()` requires a `real` matrix and returns a `real` matrix. If you attempted to pass `foo1()` a `complex` matrix, `foo1()` would abort with error, the error being that a `complex` matrix was found where a `real` matrix was expected. Mata has a number of built-in functions that work like `foo1()`. Mata's `invsym()` function for obtaining the inverse of `real` symmetric matrices is an example.

Function `foo2()` requires a `complex` matrix and returns a `complex` matrix. If you attempted to pass `foo2()` a `real` matrix, `foo2()` would abort with error. In this case, the error would be that a `real` matrix was found where a `complex` matrix was expected. Mata probably has a built-in or library function or two like `foo2()`, but I cannot find them. Such functions rarely arise because, if the calculation is valid for `complex` matrices, it is valid for `real` ones, too.

Function `foo3()` is an example of a function valid for `real` and `complex` matrices. `foo3()` allows a `real` or `complex` matrix and returns a `real` or `complex` result. Most of Mata's built-in and library functions are `numeric`.

You might expect that `foo3()` returns a `real` when passed a `real` and returns a `complex` when passed a `complex`. This is a reasonable expectation, but not because it is mathematically required. It is reasonable because Mata's built-in functions work that way. Mata's `sqrt()` function, for instance, returns missing value when passed a −1 stored as a `real`, but returns 1i when passed the same value stored as a `complex`. We at StataCorp could have written `sqrt()` to work differently. It could have returned `complex` values when called with negative values. We did not write it that way because programmers working with real values stored as `real` do not want to be bothered with `complex` results. We recommend you write your own complex functions following the same style. To achieve the "`real` if `real`, `complex` if `complex`" result, `foo3()` can be written like this:

```
numeric matrix foo3(numeric matrix A)
{
        if (eltype(A)=="real")  return(foo1())
        else                    return(foo2())
}
```

We recommend that all complex functions except those for personal use be written in this style, where `foo1()` calculates a `real` result from a `real` argument and `foo2()` calculates a `complex` result from a `complex` argument.

Here is how you should use Mata's built-in and library functions. `foo3()` stands in for whatever is the true function.

1. If you have `real` A containing real values and you are fine with the calculation not being generalized, code

   ```
   real matrix   A
   real matrix   result
   result = foo3(A)
   ```

2. If you have `real` A containing real values but you want the generalized complex result, code

```
real     matrix   A
complex  matrix   result
result = foo3(C(A))
```

3. If you have `complex` A containing real values and want the ungeneralized result, code

```
complex  matrix   A
real     matrix   result
result = foo3(Re(A))
```

4. If you have `complex` A containing complex values and want the generalized result, code

```
complex  matrix   A
complex  matrix   result
result = foo3(A)
```

B.5   Functions Re(), Im(), and C()

Mata functions `Re(A)`, `Im(A)`, and `C(A)` run on `numeric` matrices, which is to say, the argument can be `real` or `complex`.

`Re(A)` returns the real part of A. If A is `complex`, `Re(A)` constructs a `real` result containing the real part of A. If A is `real`, `Re(A)` simply returns A, and thus there is no time or memory cost to using the function.

`Im(A)` returns the imaginary part of A. If A is `complex`, `Im(A)` constructs a `real` result containing the imaginary part of A. If A is `real`, `Im(A)` returns a zero result, appropriately dimensioned.

`C(A)` returns a `complex` matrix equal to A. If A is `complex`, A is returned, so there is no time or memory cost to using the function. If A is `real`, a new `complex` result is created. Function `C(A)` is the fastest way of promoting A to `complex`.

B.6   Function eltype()

`eltype(A)` returns the element type of A. For instance, `eltype(A)` returns `"string"` when A is `string`, `"real"` when A is `real`, and so on. The function was first introduced in chapter 3 and more seriously discussed in chapter 6.

In the case of `real`, `complex`, and `numeric`, `eltype(A)` returns `"real"` or `"complex"`. It never returns `"numeric"`. `numeric` is an eltype but not a storage type. Variables declared to be `numeric` are allowed to switch between being `real` and `complex`, but at any instant, they are one or the other. `eltype()` reports the storage type at the instant it is run.

C How Mata differs from C and C++

C.1 Introduction

The term C is used to refer to the C and C++ languages in what follows.

Mata and C look a lot alike, but they are different languages. Each has unique features. They have shared features. And they have features that appear to be shared but are different in subtle ways.

This third category is the main subject of this appendix.

C.2 Treatment of semicolons

C requires that statements end in semicolons, whereas semicolons are optional in Mata. Mata's semicolon rules are

1. You must place semicolons between statements that appear on the same line.

2. You may omit semicolons when statements fit on one line.

3. When coding multiple-line statements, you may break the statement between lines only in places where it will be obvious to Mata that the statement is incomplete. You may break statements in places where

   a. not all parentheses have yet been closed,

   b. not all quotes have yet been closed,

   c. the line ends in a dyadic operator, such as +, or

   d. the line ends in a comma.

You may alternatively break the statement anywhere, just as in C, if you place Stata's and Mata's line-continuation marker, which is three forward slashes. Thus, you can code

```
y = x         ///
    + 2
```

C.3   Nested comments

Mata provides both // comments and /* comments */.

/* comments */ may be nested in Mata.

C.4   Argument passing

In C, arguments are passed by value. Consider function foo() to which a is passed. In C, a new object containing the contents of a is created, and it is the new object that is passed to foo(). Thus, if foo() changes the contents of the argument, the contents of a nonetheless remain unchanged.

In Mata, arguments are passed by address (reference). In Mata, if a is passed to foo(), it is a itself that is passed. If foo() changes the contents, a is changed.

See section 8.2.

C.5   Strings are not arrays of characters

Strings are single objects in Mata, not an array of characters. Strings may contain zero or more characters, and the length of the strings can change over the life of a variable:

```
s = "short"
.
.
.
s = "a longer string replaces the shorter string"
```

You "subscript" Mata's strings using Mata's `substr()` function.

Strings do not end in \0 in Mata. Mata's strings may contain \0, but if they do, it is a character just like any other character.

C.6   Pointers

C.6.1   Pointers to existing objects

`p = &o` works the same way in Mata and C.

C.6.2   Pointers to new objects, allocation of memory

1. Mata programmers do not explicitly allocate memory for the new objects. If `foo()` returns an object, Mata programmers can code

    ```
    p = &(foo())
    ```

    This single statement allocates the memory to contain the new object, copies the object returned by `foo()` into it, and returns the address of the newly allocated memory.

    The line above is equivalent to the C code

    ```
    p = malloc(sizeof(foo()))
    *p = foo()
    ```

2. In cases where you want to create new real, complex, string, or pointer objects, Mata's `J()` function is often used:

Code	Creates
`p = &(J(`r`, `c`, 0))`	new $r \times c$ real "matrix"
`p = &(J(`r`, `c`, 0i))`	new $r \times c$ complex "matrix"
`p = &(J(`r`, `c`, ""))`	new $r \times c$ string "matrix"
`p = &(J(`r`, `c`, NULL))`	new $r \times c$ pointer "matrix"

Use whatever values for r and c you wish, even 0.
0×0 objects do not have `NULL` addresses.

For structures and classes, the structure or class creator function is typically used. For `struct st` and `class cl`,

Code	Creates
p = &(st())	new `struct st` scalar
p = &(st(c))	new $1 \times c$ `struct st` "vector"
p = &(st(r, c))	new $r \times c$ `struct st` "matrix"
p = &(cl())	new `class cl` scalar
p = &(st(c))	new $1 \times c$ `class cl` "vector"
p = &(st(r, c))	new $r \times c$ `class cl` "matrix"

Use whatever values for r and c you wish, even 0.
0×0 objects do not have `NULL` addresses.

`J()` combined with `st()` and `cl()` can instead be used to construct new structure or class objects if you find that convenient:

Code	Creates
p = &(J(r, c, st()))	new $r \times c$ `struct st` "matrix"
p = &(J(r, c, cl()))	new $r \times c$ `class cl` "matrix"

`J($r$, $c$, st())` has the same effect as coding `st($r$, $c$)`.
`J($r$, $c$, cl())` has the same effect as coding `cl($r$, $c$)`.

In addition, if o is any object, `J()` can be used to create a new object duplicating o's type and size:

Code	Creates
p = &(J(r, c, o))	new object of type and dimension of o

3. Any function returning the desired object can be used in place of `J()`, `st()`, and `cl()`. If you want to create a new object containing a 4×4 identity matrix, for instance, you can code

    ```
    *p = & (I(4))
    ```

4. Any expression returning the desired object can be used in place of `J()`, `st()`, and `cl()`. If you want to create a new object containing a column vector of regression coefficients, for instance, you can code

    ```
    *p = & ( invsym(X´X) *X´y )
    ```

C.6.3   The size and even type of the object may change

Once you have created a new object, its size and even type can change. For instance, say you created a new 3×3 matrix and stored its address in p:

```
p = &(J(3, 3 0))
```

You can later change *p to contain a different matrix, such as

```
*p = J(50, 50, 0)
```

You could even change *p to contain values of a different type:

```
*p = "now it´s a string scalar"
```

C.6.4   Pointers to new objects, freeing of memory

1. New objects are automatically freed. New objects continue to exist as long as a pointer or pointers contain their addresses. They cease to exist (are freed) when no pointers containing their addresses remain.

2. Objects can be freed earlier than they otherwise would be by setting the pointer or pointers pointing to the object to NULL or to the addresses of other objects.

C.6.5   Pointers to subscripted values

Mata's subscripts are operators. C's subscripts are language elements.

One implication of this is that you can apply subscripts to the results of functions. Just as

```
b = v[2]
```

stores in b the second element of the vector v,

```
b = foo(...)[2]
```

stores in b the second element of the vector returned by foo(...).

C programmers expect that

```
p = &(v[2])
```

will place the address of v[2] in p, meaning that *p and v[2] will be synonyms. Changing one value changes the other.

In Mata, the statement results in the creation of a new object *p containing v[2]. *p equals v[2] just as C programmers would expect, but changing the value stored in one does not change the other.

C.6.6   Pointer arithmetic is not allowed

If p is a pointer, Mata does not allow calculation of, for instance, p+2. Mata treats the expression as a compile-time error. If p points to a vector or matrix, you cannot use pointer arithmetic to step across the elements.

C.7   Lack of switch/case statements

Mata does not currently provide the `switch` and `case` constructs that C provides. It is widely recognized around StataCorp that this is a shortcoming that ought to be addressed.

The issue is not the inconvenience of using `if` and `else if` in place of `switch` and `case`, it is the remarkable speed of execution that `switch` and `case` provide. Thus, there are two workarounds.

The first is simply to use `if` and `else if` when there are only a few dozen cases.

When there are lots of cases, the alternative is to create a vector of function addresses, such as

```
v = (&foo1(), &foo2(), ...)
```

and then call subroutine `(*v[caseno])(...)`.

C.8   Mata code aborts with error when C would crash

Mata aborts with error and presents a traceback log in situations where C code would crash.

D Three-dimensional arrays (advanced use of pointers)

D.1 Introduction

In section 18.4.2.2, we discussed how to create pointers that are not synonyms to existing variables, namely,

```
p = & (expr)
```

We used this approach to create a matrix with a ragged right edge.

You can also use this approach to create three-dimensional arrays.

D.2 Creating three-dimensional arrays

To create an $I \times J \times K$ array containing 0, create an $I \times 1$ pointer vector p and fill in each of its members with the address of a $J \times K$ matrix containing 0. The code is

```
p = J(I, 1, NULL)
for (i=1; i<=I; i++) p[i] = &(J(J, K, 0))
```

Thereafter, the matrix's elements are $(*p[i])[j,k]$.

References

Baum, C. F. 2016. *An Introduction to Stata Programming*. 2nd ed. College Station, TX: Stata Press.

Garbow, B. S. 1974. EISPACK—A package of matrix eigensystem routines. *Computer Physics Communications* 7: 179–184.

Kernighan, B. H., and D. M. Ritchie. 1978. *The C Programming Language*. Upper Saddle River, NJ: Prentice–Hall.

Nash, J. C. 1979. *Compact Numerical Methods for Computers: Linear Algebra and Function Minimization*. New York: Adam Hilger.

Author index

Subject index